LabWorX 1

Mastering the I²C Bus

Vincent Himpe

Elektor International Media BV
www.elektor.com

Published by
Elektor International Media BV
P.O. Box 11
6114 ZG Susteren
The Netherlands

Copyright © 2011 Elektor International Media BV

All rights reserved. No part of this book may be reproduced in any material form, including photocopying, or storing in any medium by electronic means and whether or not transiently or incidentally to some other use of this publication, without the written permission of the copyright holder except in accordance with the provisions of the Copyright, Designs and Patents Act 1988 or under the terms of a licence issued by the Copyright Licensing Agency Ltd, 90 Tottenham Court Road, London, England W1P 9HE. Applications for the copyright holder's written permission to reproduce any part of this publication should be addressed to the publishers.

The publishers have used their best efforts in ensuring the correctness of the information contained in this book. They do not assume, and hereby disclaim, any liability to any party for any loss or damage caused by errors or omissions in this book, whether such errors or omissions result from negligence, accident or any other cause.

British Library Cataloguing in Publication Data
A catalogue record for this book is available from the British Library

ISBN 978-0-905705-98-9

First published in the United Kingdom 2011, second print March 2012
Cover design: Helfrich Ontwerpbureau, Deventer, the Netherlands
Printed in the Netherlands by Wilco, Amersfoort

109033-UK

LABWORX

LabWorX is a collection of books, each handling a particular topic in the field of electronics.

Each volume condenses all the information, applications and notes collected during hands-on work with the covered technology, all into one book. Besides the basics of the technology, in-depth examples and applications are given as well as troubleshooting tips on how to proceed if the initial implementation fails.

The books provide a centralised repository of knowledge, each handling a particular technology. Each book may have optional circuit boards that let the reader try out the material covered. Also a website provides up-to-date information and software code examples where applicable.

This will make your life easier when dealing with the subject at hand. It is literally going from our lab to your brain so that you too can master the subject.

ABOUT THE AUTHOR

Vincent Himpe got struck by the electronics fever at the age of 8 when he burned his finger on a hot transistor. A 100 in 1 electronics exploration box pushed him onto the slippery slope of electronics technology. After graduating with a specialty in electronics he got onboard at a Semiconductor manufacturing plant as a cleanroom technician. After a couple of successful automation projects the move to the design lab quickly followed. As a design support engineer the job was to first test and debug prototype devices at the silicon level, often requiring the building of custom test equipment. After working for 10 years on both industrial and telecom devices, amongst which were the first ADSL and Bluetooth chipsets, he bid his old stomping ground farewell in 2005 and moved on to San Jose, in the heart of Silicon Valley, to work on advanced controller devices for hard disk drives.

As a seasoned electronics engineer, he holds a range of patents and has had works published in multiple leading journals and convention papers such as Imeko, TSCC, The Journal of Computer Standards and Interfaces, EDN and of course Elektuur/Elektor.

When not playing around with the latest integrated circuit or some other electronics component he can often be found on a tropical reef snapping pictures and shooting video of the underwater fauna and flora while scuba diving.

Table of contents

CONTENTS

1. Historical background of I2C 16
2. I2C from a hardware perspective 18
3. Bus Architecture 22
   3.1. Basic Terminology 23
4. The Basic I2C Protocol 24
   4.1. Flowchart 25
5. Bus Events 26
   5.1.1. Additional Terminology 26
   5.2. Idle Bus 26
   5.3. START And STOP EVENTS 27
       5.3.1. START 27
       5.3.2. STOP 27
   5.4. Putting Information On The Bus 28
   5.5. Addressing A Slave 29
       5.5.1. The Addressing Debacle 30
       5.5.2. The RULE 30
   5.6. ACKNOWLEDGE 31
   5.7. nACK or NOT ACK 31
   5.8. Key Elements To Remember 32
   5.9. Bus Stalling 32
   5.10. Getting Acknowledge From The MASTER's Perspective 33
   5.11. Giving Acknowledge From The MASTER's Perspective 33

5.12. Giving not ACK From The MASTER's Perspective 33

6. Exchanging Information 35

    6.1. Writing One Byte To A Slave 35

    6.2. Writing More Than One Byte To A Slave 35

    6.3. Reading One Byte From A Slave 36

    6.4. Reading Multiple Bytes From A Slave 37

    6.5. Determining The Slave Access Mode 37

    6.6. The Combined Data Format 38

7. Multi Master Communication 40

    7.1. An Example 41

    7.2. Additional Testing 41

8. Bus Synchronisation 42

    8.1. An Example 43

9. Special Addresses And Exceptions 44

    9.1. General Call Address 44

        9.1.1. Configuration Call 44

        9.1.2. Interrupt Call 45

    9.2. Start Byte 45

    9.3. CBUS Address 45

    9.4. High speed Master Code 46

    9.5. 10-bit Call Address 46

    9.6. Device ID 47

10. Speed Modes 47

    10.1. Standard Mode 47

    10.2. Fast Mode 47

10.3. Fast Mode + ... 48

10.4. High Speed Mode 48

11. Additional I2C USES 50

11.1. SMbus 50

11.1.1. Electrical Differences 50

11.1.2. Speed and Timing 50

11.1.3. Software Layer 51

11.1.4. Application 51

11.2. PMBUS 51

11.2.1. Applications 52

11.3. IPMI 52

11.4. ATCA 52

12. Electrical specifications of the I2C Bus 53

12.1. Enhanced I2C (FAST Mode) 53

12.2. Extended Addressing 54

13. Designing the bus system 55

13.1. Adressing 55

13.2. Bus Expansion 55

14. F.A.Q.: Hardware 56

14.1. What is the maximum aLlowed length of the bus? 56

14.2. I want to transport I2C over a long distance or off board 57

14.3. I want to extend it "by the book". Is there something like a Buffer for I2C? 59

14.3.1. NXP P87B715 59

14.3.2. NXP P82B96 61

14.3.3. NXP PCA9600 62

- 14.3.4. Hendon 5501 / 5502 63
- 14.4. Can i isolate an I2C bus? (using optocoupler or some other means) 63
  - 14.4.1. ADUM1250 / ADUM1251 / ADUM2250 / ADUM2251 63
  - 14.4.2. Homebrew 64
- 14.5. What if I run out of addresses? Is there an I2C router? 65
  - 14.5.1. Using Generic Multiplexers 65
  - 14.5.2. Using Dedicated I2C Multiplexers 66
- 14.6. Level shifting the I2C bus 66
- 14.7. Is there something like a standalone I2C controller? 67
- 14.8. Why does the clock line need to be bidirectional? 68

15. F.A.Q.: Protocol 69
   - 15.1. Q. How do i generate a repeated start condition after the last byte? 69
   - 15.2. Q. What if I was in receive mode, how do i create the repeat start? 69
   - 15.3. Q. Is it okay to abort an ongoing transmission any time? 70
   - 15.4. Q. Do i need to give the ack in read mode on the last byte? 70

16. Q&A SECTION: troubleshooting 71
   - 16.1. Q. Can i monitor an I2C bus in some way? 71
   - 16.2. USBee 71
   - 16.3. Q. Is there a way to test/debug I2C busses? 73
     - 16.3.1. Hardware Check 73
     - 16.3.2. Communication Check 73
     - 16.3.3. Corrupted Transport 74
     - 16.3.4. Multimaster Trouble 74
   - 16.4. Q. I want to experiment with I2C, Are there demo kits available? 74

17. An I2C driver in PseudoCode 76

18.	Debugging tools	80
18.1.	I2C Trigger Generator	80
18.2.	Checking Who Is Controlling The Bus	81
19.	I2C interfacing system for THE IBM-PC.	82
19.1.	Parallel PRinter Port Interface	82
19.1.1.	Software	83
19.1.2.	QuickBasic / PDS / POWERBasic / Visual basic Driver	83
19.2.	ACCESS Bus	90
19.3.	I2C In Your Computer	90
20.	Commonly used I2C devices	92
20.1.	I/O Expanders	92
20.1.1.	SAA1300 5-bit I/O Expander	92
20.1.2.	PCF8574 / PCF8574A 8-bit I/O expander	93
20.1.3.	PCF8575 16-bit I/O Expander	95
20.1.4.	MCP23017 MCP23018	95
20.1.5.	Non-volatile Expander	97
20.1.6.	Other Expanders	97
20.1.7.	Applications	98
20.2.	LED And LCD Display Controllers	99
20.2.1.	LCD Display Modules	101
20.2.2.	LED DISPLAY Drivers	101
20.2.3.	LED Drivers	102
20.3.	A/D And D/A Converters	105
20.3.1.	Principles of Digital to Analog Converters	105
20.3.2.	Principles of ANALOG to Digital Converters	108

20.3.3.	Things to Be Aware Of:		111
20.3.4.	Practical Converters with an I2C interface		114
20.4.	Audio And Video Circuitry		115
20.5.	EEPROM Memories		116
	20.5.1.	E2Prom Technology	119
	20.5.2.	FRAM Devices	120
20.6.	Real Time Clocks		121
20.7.	I2C Isolators		122
20.8.	Bus Multiplexers And Expanders		125
	20.8.1.	Bus Multiplexers	125
	20.8.2.	Master multiplexers	126
	20.8.3.	Roll Your Own	128
21.	LabSticks I2C Introduction		129
21.1.	Board 1		129
21.2.	Board 2		131
22.	LabStick 1-1: I2C Probe		132
22.1.	Probe Hardware		132
22.2.	Command Set		135
22.3.	Base Commands		136
22.4.	Data Transport		136
22.5.	Simple Stream Examples		137
	22.5.1.	Write a single byte to a slave device	137
	22.5.2.	Write multiple bytes to a slave DEVICE	137
	22.5.3.	Reading a byte from a slave	138
	22.5.4.	Reading Multiple Bytes from a Slave:	138

22.5.5.	Reading a byte from a Sub address using restart	139
22.6.	Support Functions	141
22.7.	Low Level Functions	142
22.8.	Assembly drawing	143
22.9.	Bill Of Materials	144
23.	LabStick 1-2: Universal Power Supply	145
23.1.	Assembly Drawing	146
23.2.	Bill Of Materials	146
23.3.	Selecting the output voltage	147
24.	LabStick 1-3: 24xxx eeprom	148
24.1.	Devices	149
24.1.1.	Up to 16Kbits	149
24.1.2.	Above 16Kbits	150
24.2.	Usage	150
24.2.1.	Devices up to and including 24C16 (8-bit memory address)	150
24.2.2.	Devices from 24C32 upwards (16-bit memory address)	152
24.3.	Pitfalls with E2PROMs	152
24.3.1.	Operating voltage and voltage range	152
24.3.2.	Page writing and page sizes	152
24.3.3.	Summary	155
24.4.	Assembly Drawing	155
24.5.	Bill Of Materials	155
25.	LabStick 1-4: PWM led controller	156
25.1.	Address	157
25.2.	Usage	157

- 25.2.1. Available registers 158
- 25.2.2. Input register 158
- 25.2.3. PCS0 and PCS1 register 159
- 25.2.4. PWM0 and PWM1 register 160
- 25.2.5. LED selector register 160
- 25.2.6. USING as input 160
- 25.3. Assembly Drawing 161
- 25.4. Bill Of Materials 161
- 26. LabStick 1-5: LCD / Keyboard user interface 162
  - 26.1. Initialization 164
    - 26.1.1. Device initialization 164
    - 26.1.2. LCD Display operation 164
  - 26.2. Keyboard Access 166
    - 26.2.1. Interrupt driven mode. 166
    - 26.2.2. Polled mode 168
  - 26.3. Circuit Board 169
  - 26.4. Assembly 172
- 27. LabStick 1-6: LM75 (A) temperature sensor 173
  - 27.1. Device Address 174
  - 27.2. Usage 174
    - 27.2.1. Pointer register: 174
    - 27.2.2. Configuration Register (01) 174
    - 27.2.3. Temperature register (00) 175
    - 27.2.4. Trip point register (Tos) (11) 175
    - 27.2.5. Hysteresis register (10) 175

- 27.3. Assembly Drawing 176
- 27.4. Bill Of Materials 176
- 28. LabStick 1-7: PCF8563 realtime clock 177
  - 28.1. Device Address 178
  - 28.2. Usage 178
    - 28.2.1. Control 1 register 178
    - 28.2.2. Control 2 register 178
    - 28.2.3. Seconds register 179
    - 28.2.4. Minutes register 180
    - 28.2.5. Hours register 180
    - 28.2.6. DAYS register 180
    - 28.2.7. Weekdays register 180
    - 28.2.8. Months register 180
    - 28.2.9. Years register 181
    - 28.2.10. Alarm Minutes register 181
    - 28.2.11. Alarm Hour register 181
    - 28.2.12. Alarm DAY register 181
    - 28.2.13. Alarm Weekday register 181
    - 28.2.14. CLK out control register 181
    - 28.2.15. Timer control register 182
    - 28.2.16. Timer register 182
  - 28.3. Alarm Operation 182
  - 28.4. Assembly Drawing 183
  - 28.5. Bill Of Materials 184
- 29. LabStick 1-8: 8-Bit Protected Output 185

29.1. Assembly 187

29.2. Bill Of Materials 189

30. LabStick 1-9: 8-bit Protected Input 190

30.1. Schematic 190

30.2. Circuit Board 192

30.3. Bill Of Materials 195

30.4. Some Application Information 196

31. LabStick 1-10: MCP4725 D/A converters 197

31.1. Device Address 198

31.2. Usage 198

31.2.1. Control byte 199

31.2.2. Data bytes 199

31.2.3. Write datagram 199

31.3. Read operations 200

31.3.1. Read datagram 200

31.4. Assembly Drawing 201

31.5. Bill Of Materials 201

32. LabStick 1-11: ADC081 / ADC101 / ADC121 A/D Converters 202

32.1. Device Address 203

32.2. Usage 204

32.2.1. Selector register 204

32.2.2. Conversion result register 204

32.2.3. Alert status register 204

32.2.4. Configuration register 205

32.2.5. Low limit (underrange) trip point 206

32.2.6.	High limit (overrange) trip point		206
32.2.7.	Alert hysteresis register		206
32.2.8.	Highest and lowest conversion registers		207
32.3.	Write Operations		207
32.3.1.	8-bit write		207
32.3.2.	16-bit write		207
32.4.	Read Operations		207
32.4.1.	Set pointer		207
32.4.2.	Retrieve 8-bit		208
32.4.3.	Retrieve 16-bit		208
32.5.	assembly drawing		208
32.6.	Bill Of Materials		208
33.	LabStick 1-12 PCF8591 Combined A/D and D/A Converter		209
33.1.	PCA9691		210
33.2.	Assembly Drawing		210
33.3.	Bill Of Materials		211
34.	LabStick 1-13: POTENTIOMETER		212
34.1.	Functional Differences		213
34.2.	Device Address		214
34.3.	Usage		214
34.3.1.	Writing the potentiometer		214
34.3.2.	Reading		215
34.4.	The 'D' Version		215
34.4.1.	Reading		215
34.5.	Assembly Drawing		215

34.6. Bill Of Materials 216

35. LabStick 1-14: PCA9544 I2C Multiplexer 217

    35.1. Device Address 218

    35.2. Usage 218

        35.2.1. Control word 218

        35.2.2. Interrupt logic 219

    35.3. Assembly Drawing 220

    35.4. Bill Of Materials 221

36. LabStick 1-15: PCF8574(a) / PCA9xxx Universal I/O expander 222

    36.1. Schematic 223

    36.2. Device Address 225

        36.2.1. PCF8574 225

        36.2.2. PCF8574A 225

    36.3. Usage 225

        36.3.1. PCF8574 / PCF8574A / MAX7328 / MAX7329 225

        36.3.2. PCA9534 227

        36.3.3. PCA9538 229

        36.3.4. PCA9554 / PCA9554A 230

        36.3.5. PCA9670 230

        36.3.6. PCA9672 231

        36.3.7. PCA9674 / PCA9674A 231

        36.3.8. PCA9500 231

    36.4. Assembly Drawing 232

    36.5. Bill Of Materials 232

37. LabStick 1-16: SAA1064 7 segment display 233

37.1.	Usage		235
	37.1.1.	Status register (READ ONLY)	235
	37.1.2.	Register select	235
	37.1.3.	Control register	235
	37.1.4.	Digit registers	236
37.2.	Assembly Drawing		237
37.3.	Bill Of Materials		238
38.	Appendix 1		239
38.1.	Part references		239
38.2.	Passive parts		239
38.3.	Semiconductors		241
39.	Appendix 2		243
39.1.	Additional Reading		243
40.	Index		244

1. HISTORICAL BACKGROUND OF I2C

The Inter-Integrated Circuit, in short IIC or I2C bus, was developed in the early 1980's by Philips Semiconductors (now NXP) in Eindhoven, the Netherlands. The original idea was to have an easy way to interconnect multiple integrated circuits and exchange control information. The original application was for consumer electronics, as the consumer segment is a very price sensitive market. Parallel interconnection busses require a high pin-count on device packages and the use of more complex, often double sided or multilayer, circuit boards. Both elements add a substantial cost factor to the end product. It is more cost effective to implement a few extra flipflops in a circuit and reduce the pin count on the package.

Traditional computer systems use byte or word wide data, address and control busses to accomplish this task. This requires lots of copper tracks on PCB's to route the interconnections, not to mention a bunch of address decoders and glue logic to connect everything.

Besides the cost factor, lots of control lines implies that the system is more susceptible to disturbances by EMC (Electro Magnetic Compatibility) and ESD (Electro Static Discharge) and will also generate more problems in the EM spectrum making it harder to pass certification for EM emissions.

The research done by Philips Labs in Eindhoven (The Netherlands) resulted first in a 3 wire communication system called C-bus, which later evolved into the 2-wire communication bus that is now called the I2C bus.

As said before, I2C is an acronym for **Inter IC** bus and the name literally explains its purpose: to provide a communication link between Integrated Circuits. Originally intended for cheap consumer products, the bus has found its way into numerous industrial and computer applications. Almost any given piece of equipment that contains electronics may have an I2C bus onboard. Modifications, both on hardware and software level have expanded the realm of applications and the I2C bus is also known under different names.

The System Management bus, or SM-bus, is used for system management in personal computers. Circuitry such as temperature sensors and voltage monitors use the I2C bus on the PC's motherboard. Even the memory modules use an on-board EEprom that contains the exact timing information for the memory module. All these devices are connected to the SM-bus.

The Display Data Channel (DDC) as implemented by the Video Electronics Standards association (VESA) in computer monitors that have a VGA, DVI or HDMI style connectors, is essentially an I2C bus used to read a stored block of information.

The PM bus allows power management in Computer systems, while the Intelligent Platform Management Interface (IPMI) allows system wide management using multiple intelligent I2C masters.

Finally, the Advanced Telecom Computing Architecture (ACTA) uses I2C to control the thermal management of large telecoms systems such as central office and switchboard systems.

Over the years, several other offspring's have emerged and died as well. The Digital Domestic Bus (D^2B) was intended for communication between audio, video and home automation components. The D2B system survives as a communication channel on SCART connectors for audio/video interconnection in Europe and in a modified form, as the Consumer Electronics Control channel in the HDMI interface.

The ACCESS bus attempted to find its way into computer peripherals like keyboards, mice, printers, monitors, etc... Access bus was very quickly found to be too slow and was replaced by the USB standard, but the knowledge gained from the ACCESS experiment lead directly to the development of USB 1.0.

The I2C BUS has been adopted by almost every semiconductor manufacturer in the world, and there are currently over a thousand different chips on the market that use the I2C bus as their principal mode of control. Devices range from digital I/O, audio, video, telecoms, memory, processors, DSP's, sensors and A/D D/A converters. Some manufacturers do not use the name I2C because of possible licensing or copyright issues. In other cases it is because the device is only 'compatible', and does not offer a full implementation of I2C.

An established software base is available that allows implementation of an I2C master on any CPU out there. A lot of microcontrollers and industrial CPUs even implement the I2C controller as a hardware block. For people that design with FPGA or CPLD chips, there are numerous implementations of the I2C protocol available, both as standard block or as a synthesizable core in Verilog or VHDL. Although the master controller is a quite complicated block, and a full I2C implementation is big, the slave logic is very simple and can easily be implemented in a few hundred cells. If you strip out the heavy elements like multi-master and bus-stalling the logic is merely a shift register, an edge detector and a simple state machine.

2. I2C FROM A HARDWARE PERSPECTIVE

Before we dig into the actual I2C protocol, it is necessary to have an understanding of what the hardware looks like. There is a fundamental property to the hardware implementation that helps in defining how the bus protocol works.

The bus physically consists of 2 active wires and a ground connection. The active wires, Serial Data (SDA) and Serial Clock (SCL), are both bidirectional, open drain pins.

An I2C transceiver cell in a device looks like this:

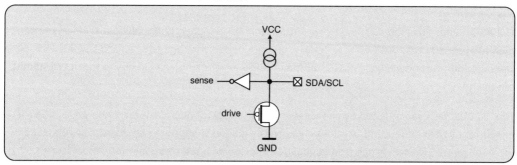

The cell is built around an input buffer and an open drain or open collector transistor. When the bus is idle, and nothing is happening, the transceiver cell is in receive mode. The output driver transistor is turned off and the signal pin (SDA or SCL) is in a high impedance state. As we are dealing with a digital control interface, it is mandatory to define a correct logic level and thus an external pull up resistor is necessary. Some I2C devices have a built in weak current source, but do not depend on these sources; it is necessary to provide your own pull up elements externally. Digital systems do not like floating signal levels, even though the inputs may have Schmitt Trigger logic.

To put something on the bus, the chip drives its output transistor, thus pulling the bus to a low level. Remember that when the bus is idle (nothing is going on) both lines are pulled high by the external pull up resistors.

If you think about it, a device can only put a logic low (0) on the bus; it cannot force its output high since it has no driving transistor on the high side. It is the job of the pull up resistor to bring the line back to a logical high level. This is an intentional design decision of the I2C bus. The detailed reasons will be explained later when we handle the protocol.

The advantages are numerous:

1) Only the low level is defined on the I2C bus. It is perfectly possible to run I2C on different signalling voltages. As long as you can pull the signal line under the maximum level for a low level, you are fine. These levels are set to 30% and 70% of the supply voltage. Some older devices use a different approach and anything below 1.5 volts is

seen as a logic low and anything above 3 volts as logic high. You need to verify the datasheets, especially if you are going to build multi voltage systems.

Any signal level above 70% of VCC or VDD is accepted as logic high. Since the pull up will limit the current, the internal ESD clamping diodes of all attached devices will automatically solve any mixed voltage problems on the high side.

2) This technique allows for some very clever software algorithms like bus mastering. If the bus is 'occupied' by a chip that is sending a 0, then all other chips lose their communications capability. More will be explained about this further on, but the gist is this: device A sends a logic one by releasing SDA. Device B simultaneous sends a logic zero. Device A loses the bus arbitration because it was not 'able' to put the logic one on the bus (someone else was pulling the signal to zero). The I2C engine must check the SDA and SCL line every time it attempts to make either signal high. If the signal remains low, the device must back off and stop talking as another device is active and has just won arbitration. Another nicety is that this mechanism not only allows for bus arbitration, but is also self healing and self error correcting: there is no data corruption for the winning device and the bus self heals after a stop has been transmitted. This will be explained in detail in the chapter on multi master communication.

However, this open collector technique has drawbacks too. If you have a long bus interconnect, the capacitive load formed by the wiring will have a serious effect on the maximum speed you can obtain. The bus capacitance, combined with the value of the pull up resistor, forms an R C network that will deteriorate the cleanliness of the signal edges. Even though the I2C devices all have Schmitt Trigger inputs this still imposes a penalty on the maximum speed you can attain.

The other problem is that the bus has relatively high impedance; this means it is more susceptible to picking up noise in the system. This can be mediated by lowering the value of the pull up resistor, but care must be taken not to surpass the maximum allowed current on the bus drivers. Check the allowed value for IoL (Output current for Low) value in the datasheets of all of the attached devices.

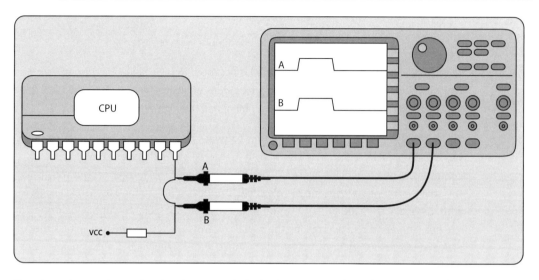

The above image shows the signal over a relatively short connection. There is no significant difference between traces A and B. This situation changes once you have long lines that have a lot of stray capacitance and that can also have other signals coupled in to them.

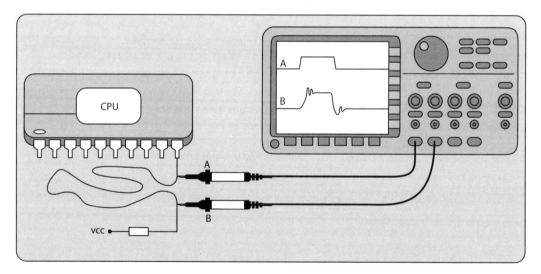

What's more is that you can get reflections at high speed. This can be so bad that 'ghost signals' can disturb your transmission and corrupt the data you transmit. In such cases, even the Schmitt-Trigger at the input of the chip will not keep you out of trouble. The reflected signals will intermodulate with the original signals and create 'hair' on the logic levels. Stray capacitance will also affect the edges of the signals and can also create ringing around the level transitions.

The following image shows you exactly what can happen.

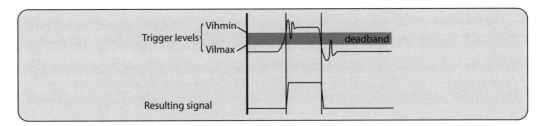

To alleviate these problems some strict electrical specs have been set forth and there are clear rules for the selection of the pull-up resistor values and the reduction of the capacitive load on the bus. For heavily loaded busses, there are now special active pull-up devices available that overcome the classic limitations of a simple pull-up resistor. These devices are essentially charge pumps. You can think of this device as a dynamic resistor. The moment the state changes it provides a large current (low dynamic resistance) to the bus. This current charges the parasitic capacitance very quickly and once the voltage has risen above a certain level, the high current source cuts out and the output current of the active terminator drops sharply so as not to overload the outputs of the individual devices. The extra current pump kicks in as long as the desired voltage has not been reached. Once the high current source disengages, there is only the static current source to maintain the appropriate logic levels.

For longer buses there are current mode and high voltage transceiver cells available, even circuits that offer full galvanic isolation are also on the market. You can also make such a circuit yourself and I will deal with these advanced circuits later on in the practical section.

3. BUS ARCHITECTURE

Now that we have established the functional working of the bus driver, we need to take a look at the interconnection level. The basic bus topology is a parallel network and each and every device connected to the network connects to the same two wires in parallel. All the SCL lines and all the SDA lines are tied together.

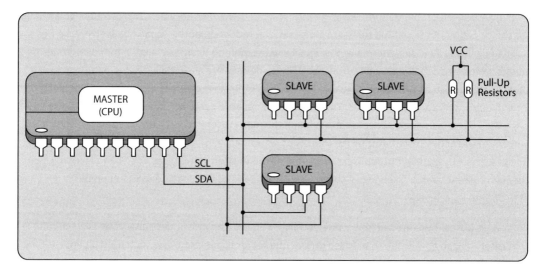

Each of these lines has an appropriate pull up resistor.

Messages are sent between a master device, typically a CPU of some sort, and a slave device. At any given point in time there is only one active bus master. There can be multiple bus masters, but only one can be active at a time. The I2C protocol provides a means to avoid collisions and has built in error correction.

The bus is point to point and even though there is a multicast address available, very few devices actually use it. Since there are multiple devices on the bus, a mechanism is needed to allow discrimination between devices. Therefore, each device on the bus has a unique address assigned to it. This address is transmitted at the beginning of a bus transaction. Only the device whose address matches the transmitted address will respond. The others will fall back in an idle mode until a new transaction takes place.

The BUS MASTER is the IC that initiates a data transfer on the bus. At that time all the other chips are regarded as bus slaves. Even in a multi master environment the other potential masters are at that time considered as 'slaves'. In any given exchange of information between the master and the slave each of them can be talking or listening. The master initiates the transport by talking and the slave initially listens. When information is requested from a slave the slave will start talking and the master will listen at that point. The master does retain control over the bus as it is still driving the

SCL line. Even though the slave is talking on SDA it needs to do this under control of the master since the master controls the SCL line.

3.1. BASIC TERMINOLOGY

In order to make life easy, here is the basic terminology as used in the I2C bus operations:

Keyword	Explanation
SCL	Serial clock line. This signal is under the control of the ACTIVE MASTER. Some datasheets call this SCK, SCLOCK or SCLK
SDA	Serial data line. This signal is primarily under the control by the ACTIVE MASTER. The ACTIVE MASTER can temporarily give control of this signal to the slave. Some datasheets call this signal SDAT, SDTA or SDATA
MASTER	A device capable of initiating a transaction on the bus and controlling the clock line.
ACTIVE MASTER	The MASTER that is currently performing a transaction on the bus.
SLAVE	A device that is not capable of initiating bus transactions. It is passive until talked to by a MASTER and it will listen or respond depending on what the MASTER instructs.
BUS Transaction	A chain of operations on the bus between a MASTER and a SLAVE. A transaction is initiated and concluded by the same MASTER (unless arbitration is lost). There are unique events on the bus that indicate the beginning and end of a transaction
MASTER TALKER	A MASTER that is sending data to a SLAVE. The MASTER controls both SCL and SDA
SLAVE TALKER	A SLAVE that is sending data to a MASTER. The SLAVE ONLY controls SDA. The SCL remains under control of the MASTER

During the remainder of this book additional terminology will also be used that will be explained when applicable.

4. THE BASIC I2C PROTOCOL

The I2C bus has evolved over the years and has gotten more and more capabilities. The current specification is a lively document that is not always easy to understand due to all the add-ons. I will explain the basic technology, and layer the capabilities in the order they were added to the spec. This section will explain the basics of an I2C operation. Further sections will tack-on the advanced parts and build your understanding of the protocol layer by layer.

The protocol of I2C has a well-defined number of states and state transition mechanisms. Each state is unique on the bus and there is no possibility for confusion. The order in the chain of events is fixed. To initiate a transaction, a START condition is generated on the bus by the BUS MASTER. All the slave devices will activate and start listening.

Once a START condition has occurred on the bus, the current MASTER will transmit the address of the slave it wants to connect to and the mode of operation. Since all slaves are active at this point they will all receive this first packet.

Information is always transmitted in 8-bit long bursts. The address field is no exception to that rule. Generally this address is 7 bits followed by a direction flag (the so called read/write bit). The I2C protocol has provisions for longer addresses and in certain cases; the address can be 10 bits long. The 8 bit per packet rule still applies. Such addresses are transmitted as two 8-bit long bursts. I will come back to this later on in this book.

After receiving this first packet, each slave will compare the contents (a 7-bit address + a read/write bit) to its own preset address. Once a slave has determined that it is being addressed it will give an acknowledge (ACK) to the master during the acknowledge cycle. If no slave is present on that particular address, this acknowledge will not happen and the master will terminate the transmission. The slaves for which the address does not match will go into an idle mode and wait for the end of the transmission signal. They will not listen to anything that is going on with the bus until this end condition received.

When acknowledge has been received by the master it will either transmit more data bytes (in case of a write), or the slave will transmit a data byte back to the master (in case of a read request). Each transmission needs to be followed by an ACK cycle if more data should follow. In the case where the master is transmitting to the slave, the slave will respond by giving the ACK. When a slave is transmitting, it is up to the master to generate the ACK.

When a transaction is complete, a STOP condition is put on the bus and all the attached devices react to this STOP by going into a sleep mode to wait for a new START condition.

It is possible to terminate a transaction by issuing a new start condition. This is done in multi master environments where a master does not want to release control over the bus. Only a STOP condition truly signals the end of the transaction. I will come back to this once we have dealt with the advanced protocol operations.

4.1. FLOWCHART

The above explanation is shown as a state diagram below.

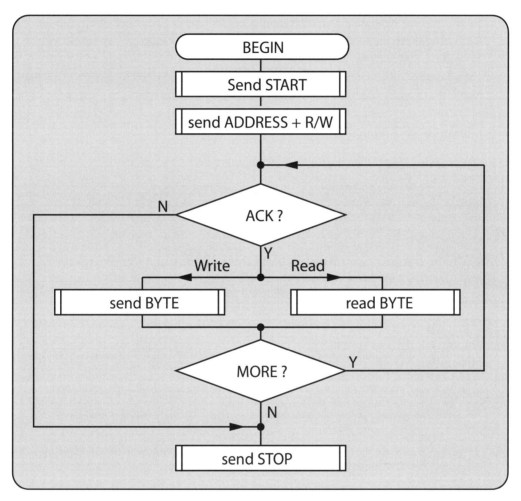

We have had several states on the BUS: START, ADDRESS, ACKNOWLEDGE, DATA, STOP. These are all unique conditions on the BUS., and we need to take a closer look at each of these in order to understand how they physically work.

5. BUS EVENTS

When reading this chapter you must keep the following things in mind:

- A MASTER is the device that initiates a message. The MASTER controls the SCL line. Later on we will see that there are possible exceptions, but for now just assume that this is true.
- A SLAVE listens to the MASTER and will send data only when requested and the slave only controls the SDA line after such a request.
- The SDA and SCL lines can only be PULLED low. They cannot be DRIVEN high. To make them high the device just releases the line. The external pull-up resistor does the rest of the work.

5.1.1. Additional Terminology

Here is a first update to the terminology list.

Keyword	Explanation
START	A bus operation that signals to all the devices on the bus that a new transaction is commencing.
STOP	A bus operation that signals to all the devices that the current talker has finished and the bus is now free.
MASTER LISTENER	The master that is controlling the SCL and listening to SDA during a transaction.
SLAVE LISTENER	Any device that can read data from SDA line, but has no control over the SCL line.
ACKNOWLEDGE	The signalling by a LISTENER, to the TALKER, that it has received a byte. This occurs during an acknowledge cycle.
nACK or no ACK	This is the failure of an ACKNOWLEDGE. The LISTENER does not respond to the TALKER or it simply does not exist.
GIVING ACK	The condition where the MASTER gives an ACKNOWLEDGE to a SLAVE.
IDLE	A state that signifies that there are no more transactions going on. After a STOP, both lines are back in the logic high state.

5.2. IDLE BUS

When the bus is in idle mode both SDA and SCL lines are pulled to logic High by the pull up resistors on the bus.

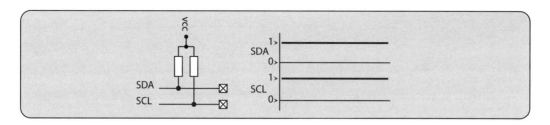

5.3. START AND STOP EVENTS.

The START and STOP events are the most basic operations you can perform on the I2C bus. Their purpose is to signal the beginning and end of an operation on the bus.

The START event acts as a signal to all connected IC's that something is about to be transmitted on the BUS. All the devices on the bus will go into listening mode and check the incoming data for a match against their address on the bus. The STOP event tells the connected chips that the message has been completed and that the bus is now free to use.

5.3.1. START

A START event begins when the bus is in idle, that is, both SDA and SCL are high. The MASTER issuing the START event (the MASTER TALKER) first pulls the SDA (data) line low, waits a moment, and then pulls the SCL (clock) line low.

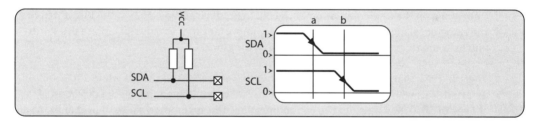

This event acts as an attention signal to all the attached devices. Any slave device, or other master that can act as a slave device, will now start listening.

5.3.2. STOP

When all operations are finished on the bus and the bus state is to go IDLE, the STOP event needs to be generated. This is a mirror in time of the START condition

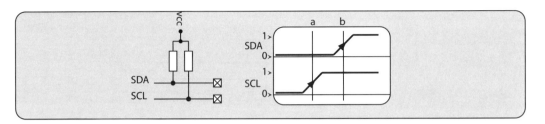

The Bus MASTER TALKER first releases the SCL line, waits a moment and then releases the SDA line. All slave devices on the bus now begin listening for a new START event. A master device can now grab the bus and issue a transaction by sending a START event.

The START and STOP events are the only ones where the SDA line changes state while SCL line is at logic high. For any other bus event the SDA must be stable BEFORE the clock is asserted (SCL going HIGH), and remain STABLE until the clock is de-asserted (SCL returning low).

5.4. PUTTING INFORMATION ON THE BUS

All information is transmitted as 8-bit (one byte) packets. The data and clock polarity are active high and the clock is valid on the rising edge. Data must be stable before the rising edge of the clock and remain stable during the high period of the clock. After the falling edge of the SCL line, the SDA line can change state. If SDA changed its state during a clock high period this would signal a START or STOP event.

> Depending on the implementation of the I2C logic, devices may latch-in the bus state on the rising or falling edge. Therefore it is very important that you do not violate the setup and hold times, even after the falling edge of the clock.

> **Note** There is a risk that delays or skew on the bus will cause SDA to change while SCL is still asserted. This will inject either a START or a STOP event (depending on the direction of change of SDA), and mess up the communication. This is common cause of problems. At all times you must maintain SDA stable for the minimum time before a rising edge, as well as after the falling edge of SCL.

Data is always transmitted MSB (Most Significant Bit) first.

Putting a byte of any kind on the bus looks like this:

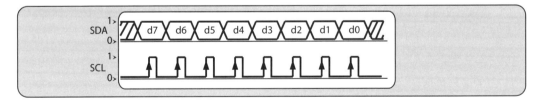

First the MASTER sets the data line to the appropriate level by pulling or not pulling the SDA line low. Then it releases the SCL line for a period of time and then pulls it low again. Now it can change the state of the SDA to the level required for the next bit and the process continues all over again.

Using the above information, a transfer could look like this:

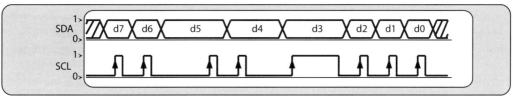

As you can see in the above waveform, it is not necessary to have a clock with a constant duty cycle. The BUS is very relaxed even to allowing that you can actually stop the clock in the middle of a transaction and then continue later on. This is very useful when the CPU has some other time critical tasks running.

Consider the following:

Your CPU is in the middle of a transaction and it gets an interrupt. It can process the interrupt first and then continue its bus transaction later on without any problems. (Try doing that on an RS232 or any other time locked bus). Since there is no minimum clock speed set you can have the communication running at whatever speed you can handle. All state changes happen while under the control of the SCL line which, by itself, is being controlled by the MASTER.

5.5. ADDRESSING A SLAVE

We already know that EVERY byte put on the bus MUST be 8-bits long. (8 clock pulses). This rule is also valid for addressing a slave. In fact the address is just a byte like any other, except for the fact that it is the first one transmitted after a START or repeat START operation.

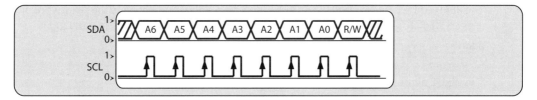

The address field consists of a 7-bit ADDRESS followed by a one bit Read/Write bit. I2C has provisions for longer address fields (10-bit address fields). However, such a 10-bit address is transmitted as two consecutive 8-bit addresses. This 10-bit address mode is handled in detail in a later chapter.

You can think of the R/W bit as an address bit. Each device has a read address and a write address. The WRITE address is always EVEN (R/W = 0), the READ address is always odd (R/W=1).

5.5.1. The Addressing Debacle

There is a lot of confusion and ambiguity about device addresses. Different manufacturers maintain different definitions and even for documentation from the same manufacturer, there may be differences depending on who wrote the datasheet in question.

The confusion stems from the fact that the ADDRESS is a 7-bit number that sits LEFT aligned in an 8-bit byte. If I tell you that the address is 0x40: is this assuming we are talking about a 7-bit word (in which case the 7 bits look like 100_0000), or do we talk about the whole byte including the R/W bit?

Do I need to shift it or not? If you tell me the device address is 0x40, I may assume this is the WRITE address since it is even. For reading the address would be 0x41. So which is it?

Here are some cases to show you just how you can spot what is actually intended.

Any ODD address means we are talking about a 7-bit address. It is uncommon to use odd numbers for an 8-bit number since this would imply READ mode (remember the LSB is the R/W bit).

Any address above 0x80 implies a full 8-bit field is given. (It is impossible to store 0x80 in a 7-bit field, the maximum is 0x7F).

The mess becomes even larger when dealing with devices that have more than one address like certain EEprom devices and I/O expanders. So it is time to invoke 'The Rule'.

5.5.2. The RULE

Here is the rule that I Use:

Devices have 3 addresses:

- The DEVICE address is the 7-bit address.
- The READ address and/or WRITE address is an 8-bit number that contains the read/write bit. A READ address is ALWAYS odd while a WRITE address is always EVEN. To find the READ address, simply add one to the WRITE address.
- The WRITE address will also be known as the BASE address.

Any code that I write will define constants for the BASE addresses. It is up to the parser to set the R/W bit to 1 if reading is required. It is faster to perform a single logic OR to set the last bit, than to perform a bitwise shift followed by a conditional OR operation. So, in order to reduce the number of CPU operations the full BYTE containing the DEVICE address and R/W bit is given.

In documentation, addresses should always be given in BINARY notation with leading zeros. This is the ONLY unambiguous notation.

Xxx_xxxx : a 7-bit address without R/W bit

Xxxx_xxxx: Full 8-bit address byte containing the R/W bit.

5.6. ACKNOWLEDGE

The acknowledge cycle is probably the one element where people have the most problems. Both MASTERS and SLAVES can issue an ACKNOWLEDGE, but ONLY the MASTER controls the cycle.

An ACKNOWLEDGE is issued by the device that has just RECEIVED a byte (the LISTENER).

When a slave is being addressed or it has received data, it will issue an ACKNOWLEDGE during the acknowledge cycle. In this case, the MASTER first releases the SDA line (leaves it set to a logic high level). The Slave issuing the ACKNOWLEDGE will then pull the DATA line low (Shown as the fat line on the next image).

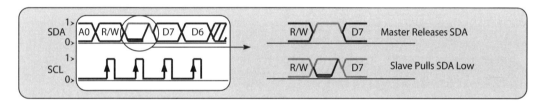

The MASTER will now send a clock pulse over the SCL line. During this clock pulse, the master will sample the state of the SDA line. If the SDA line is LOW we have an ACKNOWLEDGE. When the clock pulse has been completed, (SCL goes low again) the slave will release the DATA line.

Generally the MASTERs (mainly CPU's running software) use a timeout value. When no chip responds within a specific time, they issue a STOP and then continue with their work. This prevents your software from locking up if for some reason the addressed chip is not replying.

> **Note**
>
> A slave will start pulling the SDA line low shortly after the falling edge of the SCL signal on the 8th bit (in the above diagram the falling edge associated with the R/W bit). The slave will release the SDA on the falling edge of the clock associated with the ACKNOWLEDGE cycle. In real circuits, there are of course propagation delays.
>
> By using the falling edge of the clock, the setup time for valid data is maximized. It also eliminates the risk of the SDA changing state while SCL is rising or has already risen. This would falsely trigger a STOP condition.
>
> When 'bit banging' an I2C interface MASTER it is good practice to actually look at the SDA just prior to making the SCL line low. If the line is LOW you have a valid ACKNOWLEDGE. You will then make SCL low and MUST wait until SDA goes HIGH before continuing. This wait is required to overcome propagation delays.

5.7. NACK OR NOT ACK

In the case that a device does not pull the SDA line low during the acknowledge cycle this is denoted as a nACK or not ack. This can be caused simply because a device is not present, or a device wants to abort further operations.

5.8. KEY ELEMENTS TO REMEMBER

The MASTER is ALWAYS in control of the SCL line during the acknowledge CYCLE.

The LISTENER always gives the ACK or nACK to the TALKER. While it is always the MASTER that controls the SCL line, it is up to the LISTENER to control the SDA line during the acknowledge cycle, or not-acknowledge cycle.

Examples:

Case	Response
A MASTER has sent a byte to a SLAVE (address or data)	The SLAVE responds with ACK or nACK.
A MASTER has just received a byte from a slave	The MASTER sends an ACK if it requires another byte or a nACK to signal it is done. A nACK is followed by either a STOP or a RESTART
A SLAVE has just received a byte from a master.	The SLAVE will send an ACK if it is present, and is ready for more data.
A SLAVE has just sent a byte to a master.	The MASTER will send an ACK if it wants another byte. The MASTER will send a nACK when done. Upon receiving nACK the SLAVE will no longer send data so the master can send a stop.

The key element to remember is that the LISTENER sends the ACK or nACK to the TALKER.

5.9. BUS STALLING

Some EEPROMs have a bus stall feature. They will not release SDA after an ACKNOWLEDGE until the internal write procedure is finished. Since storing data to EEprom cells takes some time the ACKNOWLEDGE is used to indicate that the programming has been completed. After the last bit has been transferred, the EEprom starts writing the received data into its array. It leaves the SDA line in the LOW state until this action has been completed. It is the job of the master to wait until SDA becomes high before continuing. This process might seem to violate the I2C bus specification but actually, it doesn't. The I2C standard explicitly implements this mechanism.

I have once spent a whole day figuring this one out! The system occasionally did not work like it should. The second or later bytes would get corrupted. It turned out the EEPROM was still clamping the SDA line while transmission was continuing. It was a borderline valid most of the time.

Occasionally the timing would be violated so that the MSB would be corrupted. The internal delays in the slave device were sufficient to allow it to still count the clock pulse as valid.

5.10. GETTING ACKNOWLEDGE FROM THE MASTER'S PERSPECTIVE

The best way to do an ACKNOWLEDGE cycle from a master's perspective is like this:

- Set SDA line high shortly after the falling edge of SCL on the last bit transmitted. The slave will be pulling the SDA line low on this falling edge, after its internal propagation delay.
- Wait for the minimum setup time.
- Set SCL High.
- Wait for the minimum setup time.
- Sample SDA. If SDA is low we have a valid ACKNOWLEDGE. If SDA is still HIGH there is no acknowledge (nACK). You may opt to wait as long as possible using a time out value. Sample the SDA line as late as possible before the next step.
- Make SCL low
- Wait for the minimum hold time.
- Check SDA and wait for it to go back HIGH. You can add a timeout mechanism here. If the SDA line does not return high you have a stalled device, (the slave may be stalling the bus to complete its internal work. This is a valid mechanism) or the bus may be locked up, then time to activate your emergency escape.
- Proceed normally.

5.11. GIVING ACKNOWLEDGE FROM THE MASTER'S PERSPECTIVE

Here is how a master would give Acknowledge to a slave. This occurs immediately after the falling edge of SCL on the last bit.

- Set SDA low.
- Wait for the minimum clock time.
- Raise SCL
- Wait for the minimum clock time.
- Drop SCL
- Wait for the minimum clock time.
- Release SDA

5.12. GIVING NOT ACK FROM THE MASTER'S PERSPECTIVE

Here is how a master would give a nACK to a slave:

- Set SDA HIGH.
- Wait for the minimum clock time

- Raise SCL
- Wait for the minimum clock time
- Drop SCL

6. EXCHANGING INFORMATION

So far we have only seen the basic bus operations. Now it is time to add them together and perform some real communication.

6.1. WRITING ONE BYTE TO A SLAVE

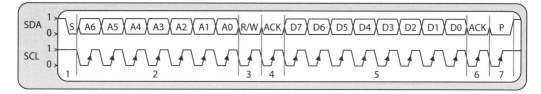

Any I2C transaction must begin with a START operation (1) and end with a STOP operation (7). The first byte transmitted, also known as the ADDRESS byte (2) must contain the slave address as well as the read write bit. Since we are dealing with a WRITE operation this R/W bit is set LOW. During the ACKNOWLEDGE cycle (4) the SLAVE will pull SDA low. The MASTER will then proceed by sending the data byte (5). The DATA byte is followed by another ACKNOWLEDGE cycle (6). Since this is also the end of the transmission, the MASTER now proceeds by generating a STOP condition on the bus (7).

6.2. WRITING MORE THAN ONE BYTE TO A SLAVE

After the SLAVE has responded with an ACKNOWLEDGE (see above) the next 8-bits are placed on the bus. Now you have to issue another ACKNOWLEDGE event and wait again for the SLAVE to ACKNOWLEDGE. If you are finished with data transmission, issue a STOP command, following this, the bus is then idle again. If you need to send more data, then simply repeat this cycle: send 8-bits followed by an ACKNOWLEDGE cycle.

There is no physical limit on how much bytes can be transmitted; it just depends on the device you are talking to. Some devices have multiple internal registers and the first byte will land in the first register, the second byte goes in the second register and so on. When you write more information than there is room in the devices the address counter typically rolls over to 0. If you have a device with 4 registers the 5th byte will land back in the first register. The 6th byte will end up in the second register and so on.

A multi-byte write looks like this:

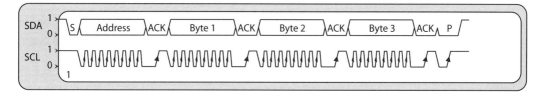

> **Note** Any byte transmitted MUST be followed by an acknowledge cycle. The listener uses the acknowledge cycle to complete the transfer of the byte and latch it into the internal registers. Should a stop event be received after the last bit has been transferred, the data will not be taken into account and the write operation will fail. This is a common mistake.

6.3. READING ONE BYTE FROM A SLAVE

Reading looks kind of the same as a byte write. The difference is the handling of the SDA line and the ACKNOWLEDGE.

The MASTER generates a START, transmits the device address and the R/W bit (set to 1) and waits for an ACKNOWLEDGE. So far this is the same as a write transaction. After the ACKNOWLEDGE cycle the MASTER will RELEASE the SDA (data) line. The SLAVE will pull the SDA line low or leave it high depending on the bit it wants to transmit. On every clock pulse that the MASTER generates, the SDA line will be in the state set by the SLAVE. When all 8-bits have been read, the MASTER will send a STOP to terminate the operation.

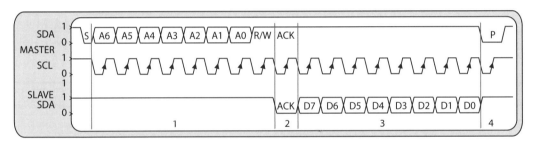

The above timing diagram shows what is going on and who controls the bus at any point in time. The transaction starts with the master sending a START condition followed by the address and R/W bit set to 1. This time the R/W bit is set to 1 because we want to perform a read operation. During the ACKNOWLEDGE phase (2) the master releases SDA (top waveform) and the SLAVE pulls it LOW (bottom waveform). The master generates a clock cycle (middle waveform). So far everything is normal. This time though, the MASTER does not take control of SDA but leaves it in its HIGH state. The SLAVE will place the SDA line in the correct state according to what information it wants to transmit (3). The MASTER is still in control of the SCL line. The SLAVE will change the state of the SDA line on the falling edge of each clock cycle and the MASTER will read the information from SDA either on the rising edge or while the clock is high.

After transmission of the 8th-bit the SLAVE will release the SDA line. Now the master will issue a STOP command (4) by making SDA low, raising the SCL and then raising the SDA. The transaction is over and one byte has been read from the slave device.

> **Note** It is also possible to send a nACK before sending a STOP command. I2C devices will generally stop the operation in this case. It is possible that devices retransmit the last byte. This is something that needs to be studied. Fixed logic devices do not do this but a software implementation could.

6.4. READING MULTIPLE BYTES FROM A SLAVE

If you give an ACKNOWLEDGE to the slave you must read another byte. Immediately after the ACKNOWLEDGE the slave will take hold of the SDA line and you will no longer be able to give a stop before clocking the next 8-bits.

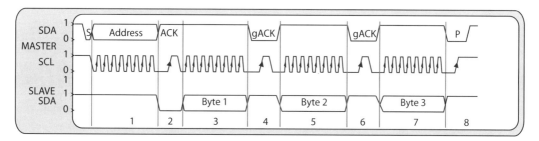

The image above shows exactly what is going on. The MASTER initiates a transaction by issuing a START condition followed by a device address (1) in read mode. The SLAVE acknowledges during the ACK cycle (2).

Since the R/W flag was set to 1 the master releases the SDA line and the slave now takes control of the SDA line. For the next 8 clocks, the SLAVE outputs the data bits onto the SDA line (3). On the falling edge of the 8the clock, the SLAVE releases SDA. Now the MASTER will GIVE ACKNOWLEDGE (gACK) by pulling SDA low and issuing a clock pulse (4). The SLAVE takes control of the SDA line and continues by transmitting the next 8-bits (5). Here again the MASTER sets GIVE ACKNOWLEDGE (gACK) to the slave (6) and the slave continues by transmitting another 8-bits (7). The MASTER now issues a STOP command and the bus returns to an idle state.

6.5. DETERMINING THE SLAVE ACCESS MODE

How does your SLAVE know whether you want to read from or write to it? That's an easy one. The READ/WRITE bit in the address is determining this. When this bit is set to 0 it means you want to write to your SLAVE. When it is set to 1 it means that you want to READ from the slave. In essence, the even addresses are WRITE addresses and the ODD addresses are READ addresses. Each device has a consecutive WRITE and READ address.

Example: a PCF8574 General Purpose 8-bit I/O port.

SLAVE address to WRITE is (01000000) b = 64d

SLAVE address to READ is (01000001) b = 65d

You can have a theoretical maximum of 128 devices on a BUS. In a practical environment though, this is not the case. The I2C specification sets aside a couple of addresses which you are not allowed to use.

Once a device address is transmitted, the access mode is set for the remainder of the transaction. You cannot change from write to read or vice versa. The current transaction must be completed using a STOP command and then a new transaction needs to commence using the new access mode.

This has potential problems. Let's assume that a device has multiple registers inside and wants to give you the capability to read a specific register rather than having to stream out all of them in sequence until you have the data you need. Examples of such devices are the E^2PROMs (EEPROM) and parts that have RAM onboard like the real time clock chips. For this, the I2C specification provides a combined data format.

6.6. THE COMBINED DATA FORMAT

Suppose you have a 128 byte memory on the bus and you want to read the 84th byte. Normally you would have to read the first 83 bytes before getting to the bits you are looking for. This takes too much time, occupies the bus and wastes valuable CPU time on the master.

Such devices typically allow you to set an internal address pointer. This pointer is the first byte, or first few bytes, transmitted after the initial device address. Once the pointer is set, subsequent operations start based on this pointer. Now there is a caveat in this technique. Remember that the access mode (R/W) is set in the address field and this mode cannot be changed during a transaction. So how do we read from a random address? We need to access the SLAVE in WRITE mode to set the pointer, STOP the transaction and access the slave again in READ mode. It looks pretty simple but has a potential pitfall.

If you are the only master on the bus accessing this particular device there is no problem. If there are multiple masters that can access this device you may be in trouble.

Suppose you have two masters that want to access this device. It is possible that MASTER 1 sets the pointer to the desired location and issues the STOP. Due to timing, MASTER 2 sneaks in and manages to change the pointer before MASTER 1 can issue a new transaction. After MASTER 2 completes his transaction, MASTER 1 now performs an operation and gets the wrong data!

The MASTERS would have to monitor each other's operations going to devices at all times. Even then they may not be out of the woods. The masters would have to analyze what is being performed to decide if they can now take their turn in accessing the slave device. This is too much trouble and takes up CPU time and code space. So, a simpler method was introduced.

Another condition potentially exists in a master code that is multi tasking. It would be possible for one thread to inject operations that would mess up an operation that was in use. Testing the STOP flag must be performed to see if the bus is free before injecting an operation. Otherwise, an exception handler is required.

The basis of a bus transaction is that the bus only goes to idle after a STOP condition occurs. As long as no STOP is transmitted, the bus is considered to be occupied. So the solution is simple: do not send a STOP command but send another START command.

The state machine that controls a slave device is made in such a way that a START condition forces the machine to begin at step 1. Even if during a transaction a START is detected, the device will fall back into the initial state.

Now it is easy to perform 'uninterruptible'accesses that changes the access mode without the risk of another master taking control in-between the changing of the access mode.

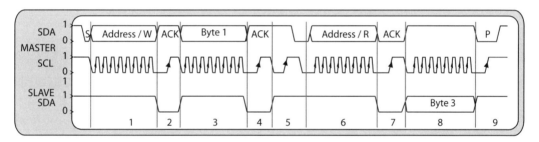

To start this kind of transaction the master simply transmits the slave address with the R/W bit cleared (1). The slave will send acknowledge as usual (2). The master then transmits one or more bytes and waits for acknowledge after each of these bytes has been sent (3 and 4). After the last byte has been written and acknowledge has been received (4) the master now executes a repeated start (5). The repeated start makes the SDA line high, then makes the SCL line high and pulls the SDA line back low while SCL is kept high. As a last step the SCL is brought back low.

All the devices on the bus have now returned to listening mode. It is most logical that you will now transmit the slave address again but this time with the R/W bit set to 1 (6). However, it is also possible to talk to a completely different device as well!

The device will respond as normal by giving acknowledge (7) and then transmitting the first byte of information. You can read as many bytes as you want and terminate with a stop operation (9).

It is also possible to not terminate with a stop operation but send another START operation instead. This will reactivate all devices back into listening mode and you can now change the access mode, or select a different device to talk to.

The advantages of the repeated start mechanism go beyond assuring uninterrupted access to a device. If you have 2 CPU's on your bus which may want to take the bus, this will assure you that you will be able to continue your operations on the bus without interference from the other CPU.

Remember that when you have generated a START and have sent the SLAVE address, the other CPU will be waiting until a STOP appears on the bus. So he will not try to put anything on the bus.

The drawback is that this potentially blocks the bus for a long time. The other masters may be stalled waiting for the bus to become free.

If you do not have a multi master environment you don't need the repeated start mechanism and you may be better off implementing simple read and write operations as it makes your code base easier and smaller.

7. MULTI MASTER COMMUNICATION

So far we have only dealt with the basic I2C operations. The bus has a lot more going for it than that. I mentioned a couple of times that you can have more than one master on the bus. There can only be one ACTIVE master and care must be taken that one master does not interfere with the other. One part of the solution is to use the restart mechanism to block access to a device in order to avoid conflicts. This solves potential problems during a transaction.

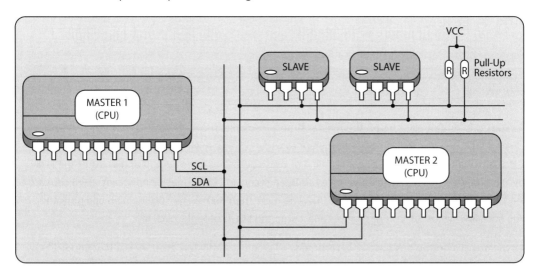

The question is: how do you avoid collisions when both masters attempt to take control of an idle bus?

Let's assume the bus is free (SDA and SCL are both logic high and a STOP condition was the last event on the bus). Both master 1 and master2 have operations pending. The question is who goes first? There is no connection between the masters other than the SDA and SCL lines.

The key problem is that there is no synchronization between the masters. The beauty of the I2C bus is that there doesn't need to be.

Remember that a device can only make a line LOW. It is the job of the pull up resistors to return the line to a logic high state when nobody is pulling a line low. This is the key to solving the problem.

When you (as a MASTER) change the state of a line to HIGH, you MUST always check that it has gone to the HIGH level. If it does not return HIGH, then someone else is pulling the line LOW. Of course when you are in an acknowledge cycle this is normal behaviour. But outside of an acknowledge cycle it means the bus is occupied and you should back off and wait for a STOP condition to occur before trying again.

So now we have our detection mechanism that allows us to determine when someone else is talking, but what about error correction? The answer is simple: it is not needed. The device that was unsuccessful in setting the lines high will release both lines and wait for a STOP event to occur on the bus. The other device has not noticed anything. It wanted to set the line low and the line went low. There is no data corruption. This device just won arbitration without even knowing it.

In short, the device that wins the arbitration continues talking and is not aware that something has happened. No data got corrupted. The losing device simply stops bus operations and waits for a STOP event.

7.1. AN EXAMPLE

In this example both masters start transmitting simultaneously.

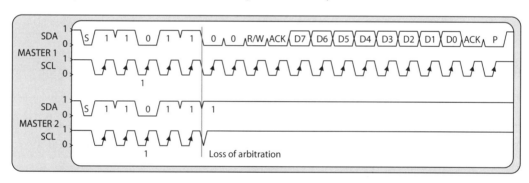

Initially all goes well and the two devices are roughly in sync with each other.

At a given point, Master 1 places a logic 0 on the SDA line. Master 2 decides to release the SDA because it wants to transmit a logic 1. The implied rule is that you always check that the line does indeed go high when you release it. Since this not the case, master 2 loses the arbitration on the bus and backs off by immediately releasing SCL. This does not corrupt the clock line as master 1 is still keeping it low. SDA in this case was already released since master 2's intention was to send a logic 1.

Since master 2 has lost arbitration it will now not take any action on the bus until a STOP condition is encountered.

7.2. ADDITIONAL TESTING

In an ideal world, MASTER 2 should not just switch into back off mode. If the arbitration loss occurred during the sending of the address, it should switch to receive mode and catch the remainder of the transmission. After all it is possible that MASTER 1 is trying to talk to MASTER 2.

In a multi master environment, this is possible. For the above example, assume we have 4 masters for which the last 2-bits identify their own address. It is possible that one of the other masters has actually initiated a transaction to talk to MASTER2 at exactly the moment master2 was trying to send something as well.

A good software implementation would thus pick up the remainder of the transmitted information and decide what to do. This complicates the software tremendously and therefore it is easier to simply back off and leave it to the other master to retry its operation. After all, if master 2 is in back off mode it will not give an ACK to master 1. That one will decide that the device is busy, issue a STOP condition and try again later.

If you have hardware assisted I2C it becomes easier to implement these features. In a bit banging engine this gets pretty hairy to implement correctly.

From the above story, we can conclude that it is the device that is pulling SDA LOW that wins. The master that wanted the line to be HIGH, when it is being pulled low by the other, loses access to the BUS. We call this a 'loss of arbitration'. When arbitration is lost, the losing entity must release SDA and SCL and wait for a STOP condition to appear on the bus.

The above example showed a situation where the two CPU's were in perfect sync with each other. In most situations, this will not be the case. But even then the arbitration will still work. Suppose one of the CPU's missed the START condition and still thinks nothing is going on, or it came just out of reset and wants to start talking on the bus. These are real life cases that WILL happen in a multi master environment. (Remember Murphy's Law!). Consider a multi master system that has just been powered up. You will not know which CPU will come out of reset first.

Not only does the I2C bus have inherent arbitration, it is self-healing. No information is corrupted during an arbitration event. The message from the winning CPU is not distorted in any way. The Masters must monitor both SDA and SCL. Whenever they fail to raise either SDA or SCL when they should, this creates a loss of arbitration event. There is one exception though, should this occur during an acknowledge cycle there may be something else going on.

8. BUS SYNCHRONISATION

The I2C protocol also includes a synchronization mechanism. This can be used between masters. However to my knowledge there are currently no chips that use this mechanism. There may be software implemented slave devices that use this mechanism, or you can use this when designing a system but this must be implemented in software.

There is a potential problem when slow reacting slaves (like software slave devices) are attached to the bus. Suppose the following: The master reads a byte from a slave, let's say an A/D converter. The slave needs some time to make a conversion. The master addresses the slave in read mode and this will initiate a conversion.

During the acknowledge phase the slave device may actually pull SCL low. The master will not be able to raise SCL at this point. The exception handler in the master will then wait for SCL to become high.

Don't confuse this with a loss of arbitration. Loss of arbitration only happens outside of an ACK cycle. This is a special case where a slave pulls down the SCL line during the acknowledge cycle. Remember the definition of a slave: A device that cannot drive the SCL line. Well, this is the exception to the rule.

8.1. AN EXAMPLE

The image below shows the timing of what is going on:

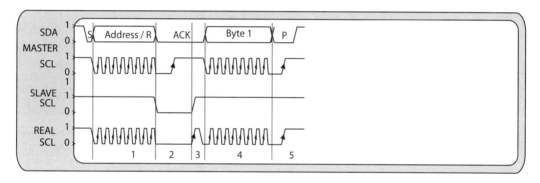

The transaction starts as usual with a start condition followed by the slave address (1). As soon as the SCL line turns low on the last bit transferred by the master, the slave pulls its SCL and SDA lines low. The master attempts to make SCL high in order to perform the ACK cycle (2) but fails. The SCL line remains low because the slave is blocking it low. Whenever the slave finishes its internal operation it releases the SCL line (3). Since the master had already made its SCL pin high the real SCL signal now jumps high. The master can now complete the acknowledge cycle and the transaction resumes.

The caveat with this mechanism is that it can cause a deadlock on the bus. Whenever the SCL line is clamped low it is not possible to perform any I2C operation. If the slave is stuck at this point the master cannot even force a STOP condition, therefore a timeout mechanism must be provided that can free the bus.

Another drawback is speed. The BUS is locked at that moment. If you have rather long delays (long conversion time in our example above) then this penalizes the total bus throughput a lot.

The example I gave was based on a software implementation. Real A/D converters do not use this type of synchronization. They start a new conversion upon the completion of transmitting the result. Therefore, if you need 'fresh' data you need to perform two consecutive read operations. The first read result is discarded, but it will have triggered the converter to sample its input and digitize it. The second read will then fetch this data and start another conversion.

The Synchronization mechanism is nothing else than a sort of handshaking between smart devices. Keep in mind that, it is possible to go into a deadlock mode. When using this synchronization mechanism you must take care to avoid all possible causes of a deadlock. This makes the interface driver very complex and will require the setting of timeout limits at various steps in the code.

9. SPECIAL ADDRESSES AND EXCEPTIONS

The I2C bus has provisions for so called 'special addresses'. Some of these are reserved for compatibility with the older CBUS. Others are reserved for future expansion.

7654	321	R/W	Function
0000	000	0	General Call address
0000	000	1	START byte
0000	001	X	CBUS address field
0000	010	X	reserved for a different bus format
0000	011	X	reserved for future purposes
0000	1XX	X	High speed master code
1111	0XX	X	10 bit slave call address
1111	1XX	X	Device ID

9.1. GENERAL CALL ADDRESS

The general call address is used mostly in a multi master environment. Some slave devices allow the master to perform a 'software' reset using this address. Very few devices implement this feature. Any device that can handle this general call address will respond with an ACKNOWLEDGE, but the master will have no idea just how many slaves actually respond.

If acknowledge is received on this address, it means that there is at least one slave capable of handling it. The master can now send additional bytes, but do keep in mind that ALL the slaves that respond to this general call address will receive and process this information.

There are two modes of operation in a general call. One is used to send configuration data to a slave, the other is used as an 'interrupt' call.

9.1.1. Configuration Call

When the LSB of the second byte is a '0' then we are dealing with a configuration call. The I2C spec has provisions for the following configuration codes:

7654	3210		Function
0000	0000	(00h)	This is an illegal code to use as second byte.
0000	0100	(04h)	Instruct it to re-read the programmable part of its SLAVE address. The SLAVE will resample its hardware ADDRESS pins.
0000	0110	(06h)	Identical as the above code but the slave will also reset to power-up condition.

Only two valid opcodes have been defined for this kind of call. Opcode 0x04 will force ALL attached slaves, which can handle such a call, to resample their hardware address pins. If you have reconfigurable hardware in your system then this call can be used to reprogram the addresses of the slaves. An I/O circuit may change the hardware pins on the devices. The configuration call 0x04 then forces the slaves to reread the new state and change their address based on the read pin states.

Opcode 06 goes beyond that by also erasing all the internal registers in a slave device to their power up values. This will NOT erase EEPROM cells. It merely resets the normal read/write registers in the slave, not the EEPROM array.

9.1.2. Interrupt Call

If the LSB of the second byte is logic '1' then we are dealing with an interrupt call. An interrupt call allows a MASTER to send a message to another device. For example a keypad encoder could send such a packet upon a key press. The seven unused bits in the second byte will now contain the address of the device that is currently sending. Subsequent bytes may contain the payload. There is no set payload length. It is up to the programmer to come up with a payload management system.

The interrupt call can only be used in a multi master environment. It is an unknown as to when such a call will be placed. All devices, including all other masters on the bus, must listen to the general call, this allows for interrupt driven messages. Since this kind of general call contains the address of the originator a reply can be sent by another device.

9.2. START BYTE

The START byte allows for synchronization between bus MASTERS. It is NOT allowed for any device to acknowledge the START byte. This mechanism helps in offloading the constant polling of a 'soft' slave or master that it may have to do when being addressed.

There are NO hardware slaves that require this mechanism.

9.3. CBUS ADDRESS

The CBUS Address effectively locks all the slaves on the bus until a STOP condition is encountered. There are NO I2C slave devices that use this as address. The CBUS differs from the I2C bus in the sense that it has a third line that is used to select the device being talked to.

In order to save I/O pins on a processor you can recycle the SDA and SCL pin to be the CLOCK and DATA pin of the CBUS system. The problem is that I2C slaves may react to transmissions that are not intended for them. By sending the CBUS address all I2C slaves will be locked down. The basic bus mechanism freezes a slave that does not have a matching address bit until a valid STOP condition occurs. By sending the CBUS address ALL slaves will fall in this category. They will all be waiting for a valid STOP. The processor can now activate the DLEN line (DLEN is a CBUS control line) and perform the CBUS transaction. When the processor is done, it deactivates DLEN and creates a STOP condition on the I2C bus. All IC devices will now resume normal I2C operation.

9.4. HIGH SPEED MASTER CODE

The HIGH speed master code allows a high speed master to switch between normal operation and high speed operation. High Speed mode allows transmission at up to 3.4 Mbits/s. This mode does not allow for clock synchronization or bus arbitration. Care should be taken that no other devices can start talking. Since there is no arbitration the data would be corrupted and no error detection is possible. The physical I2C hardware is modified as well in this mode. There is a separate chapter on High speed mode later in this book.

Eliminating the arbitration mechanism allows the speeding up of the I2C engine since the normal check to see if SCL or SDA did indeed go high when set so, can be omitted. This saves clock cycles and processing time.

9.5. 10-BIT CALL ADDRESS

In order to allow for more devices on the I2C bus an extended addressing scheme was developed. Instead of a 7-bit address, a 10-bit address is used. The 10-bit address is encoded as two bytes. The normal address field for I2C devices contains the 10-bit address 'command' and the 2 MSB's of the 10-bit address. The second byte contains the 8 LSB's of the 10-bit address.

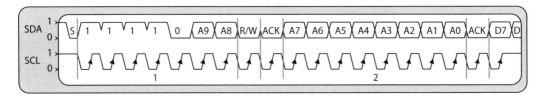

The 10-bit 'command code' has 1111_0 as the MSB's of the normal address field. The image above shows the exact layout of the 10-bit address field.

It is possible that multiple devices respond on the first ACK cycle. Only on the second ACK cycle will the device that matches all 10-bits respond. In case your system uses the 10-bit addressing in conjunction with 7-bit addressing you need to make provisions to detect both ACK signals to obtain a valid 10-bit ACKNOWLEDGE from a 10-bit slave.

Normal I2C devices will not use addresses in the 1111xxx range. Even though the combination 1111_0 is only used for 10-bit addressing, the 1111_1 address header is reserved for other purposes.

9.6. DEVICE ID

Device identification is a new feature that very few devices actually support. This special address allows the requesting of information from an attached device. The master must transmit this special address with the R/W bit cleared, followed by the address of the device it wishes to identify. The master must then send a repeat start and the device id address again, but this time with the R/W bit set. The device whose address was transmitted earlier will now respond by sending out 3 bytes. These bytes have 3 unique identifiers encoded in them: a 12-bit manufacturer code, 9-bit part identification and a 3-bit revision code. The master terminates by generating a STOP command.

10. SPEED MODES

Over the years that the I2C has evolved, not only from a protocol perspective, but also with the addition of 10-bit addressing, speed improvements and other features. The emergence of I2C hardware master building blocks, and the inclusion of such blocks as standard elements in a lot of popular microcontrollers, have allowed for the bus speed to increase.

The I2C bus today has 4 distinct speed modes. Any I2C device is always downward compatible. That is, a fast device can always run at a lower clock speed.

10.1. STANDARD MODE

This is the original I2C bus mode and is limited to roughly 100 Kilobit/s transfer rates. All I2C devices are compatible with this mode.

10.2. FAST MODE

Fast mode I2C pushes the speed up to 400 Kbit/s. The protocol and electrical specification, with the exception of timing, are identical to Standard mode, but there is a couple of subtle difference that set this mode apart from the standard mode.

- A Fast mode device must float the SDA and SCL pins when it is powered off.
- The inputs of a Fast mode device have adaptive Schmitt triggers and spike suppression.
- The output drivers have slope control for the falling edges of SDA and SCL.
- The pull up resistors must be adapted to allow for a faster rise time of SDA and SCL. For heavy capacitive bus loads (more than 200pF) this pull up resistor may have to be replaced by a current source or a controlled resistor.
- Compatibility with CBUs devices is not required since these busses cannot run at these speeds anyway.
- Fast mode devices must be able to synchronize with the 400 KHz clock and they must have clock stretching capability. Clock stretching allows a SLAVE device to lock the SCL line in a low state in order to 'stall' the master.

It is inadvisable to connect standard mode devices to a fast mode bus if you are only going to run it in fast mode. The timing requirements may cause standard mode devices to misbehave and/or latch up the bus.

Running a mixed standard mode/fast mode bus always in standard mode poses no problems at all.

10.3. FAST MODE +

Fast mode+ not only boosts the speed up to 1 Mbit/s, but also allows for a larger capacitive load on the bus. Just like fast mode devices, fast mode+ devices are backward compatible from a protocol perspective. You can run fast mode+ devices at standard clock speed. Mixing Standard and fast mode devices with fast mode+ devices on the same bus is only acceptable if you are going to run the bus at the speed of the slowest device.

Fast Mode+ devices typically have stronger bus drivers and are more tolerant of the slow rise and fall times. This allows for a bus load up to 400pF without compromising speed.

Active pull up devices will be needed on this kind of implementation.

10.4. HIGH SPEED MODE

High speed mode allows speeds up to 3.4 Mbits/s but also departs from the other I2C bus modes. Even though the same protocol and data format is maintained as with the other bus speeds, arbitration and clock synchronization are not performed.

A number of modifications allow for the transmission of data at high speed.

High speed masters have an additional current source on the data line. This allows for a faster rise time of the SDA line. This source is only enabled when the master is transmitting in High speed mode.

There is no arbitration or clock synchronization during a high speed transaction. Once a high speed transaction has started the arbitration mechanism is turned off. In a multi master environment the masters will be aware of this.

The high speed masters use a 30% duty cycle for the SCL. The clock is high for 1/3 of the clock period. This allows for a stretched setup time.

It is advisable to make a separate bus that only interconnects high speed devices to avoid bus loading. Some masters have dedicated pins for high speed and other modes.

The buffers of high speed devices have additional signal conditioning in terms of spike suppression and level detection. The output drivers have slope control for the falling edge. This eliminates potential ground bounce and ringing. It is advisable to include series terminating resistors on the high speed lines.

Because of the lack of arbitration, this high speed mode requires some additional mechanisms. By default the high speed mode is turned off, even if the master is a high speed device. To initiate a high speed transfer the master must transmit the master code. This is a reserved code; it serves for blocking of other devices on the bus. This master code is transmitted in one of the non high speed modes that are still compatible with the slowest device on the bus. Also the master has arbitration turned on. At this point arbitration happens as usual. Should another device claim the bus and win, then we need to back off.

The master code is 0001xxx and allows for 8 potential devices. The high speed master will transmit its own address. Note that in this case the 8th bit, that is normally a read/write bit, doesn't matter. The 8th bit is used as a normal address bit. The four leading zeros increase the probability of a high speed master claiming the bus. The only possible arbitration fight will be between other high speed masters. The actual master address is a matter of software and the system designer is free to associate addresses at will. It is advisable though to use the base address (0000_1000) only for diagnostic purposes.

No acknowledge is given. Normal I2C devices cannot have these reserved addresses, and the other masters will not react since we are transmitting our own address.

At this point the SDA line is high (we come out of nACK with SDA high. Remember nACK is leaving SDA high and giving a clock pulse on SCL) and the high-speed master will turn on his current mode driver on SDA. The High-speed master will now make SCL high and check if any other device is attempting a clock synchronization procedure. Once the SCL is high the master will now drop SDA. This creates the repeated start condition and resets all slave devices to listen for an incoming address without the master relinquishing control of the bus.

At this point the master will turn on the current mode driver on SCL and speed up the clock. The arbitration process is turned off now. We have entered into, and remain in, high-speed mode until a stop condition occurs on the bus.

The STOP condition must be given with the timing of the slowest device on the bus. If we were to generate a stop condition with a faster timing the slow device may not detect it correctly.

Typically masters that are capable of this high speed mode will have separate SDA and SCL lines called SDAH and SCLH. It is advised to connect only high-speed capable devices on this bus. Even though the bus is backward compatible it will avoid potential trouble with slow speed devices that might enter a wrong state.

High-speed masters can have an I2C bridge that disconnects slow speed devices. For the programmer there is no distinction between the two busses. When a non high-speed transaction is ongoing, all devices get the signals. Only when high-speed mode is requested will the low speed devices be disconnected by turning off the pass transistors that connect to the low speed devices.

11. ADDITIONAL I2C USES

I2C has found its way into lots of applications beyond those it was initially intended for. Almost any modern electronics device has one or more I2C devices onboard. The ease of use, low pin-count and versatility has popularized this protocol. Some of the applications of an I2C bus system have begun to live a life of their own and have even acquired a name for themselves.

11.1. SMBUS

The System Management bus was introduced by Intel in mid 90's in their Pentium class chipsets. Today this bus is present on any computer motherboard on the market. The SM-bus is basically an I2C bus with small tweaks here and there. The protocol and capabilities are unchanged, only some electrical and timing parameters have been altered.

11.1.1. Electrical Differences

The SM-bus lowers the ViHmin (the minimum voltage that is required on an input to be detected as a logic 1) from 3 volts for I2C to 2.1 volts. This allows easier interfacing with low voltage devices. Most SM-bus networks run off a 3.3 volts power supply. The bus remains tolerant up to 5 volt, so mix and match is still not a problem

The ViLmax (maximum input voltage that is still seen as logic 0) is fixed at 0.8 volts for SM-bus devices, where on I2C this was set at 1.5 volts (on a 5 volt supply).

The bus drive strength has been beefed up; a normal I2C device can only sink current of 3 mA maximum. In order to reduce the sensitivity to noise on motherboards this current has been raised to 4mA for the SM-bus. In practice this is not a real problem; normal I2C devices will also achieve that without any problems.

The SM-bus also provides a third signal that acts as an interrupt. Whenever a device needs attention it can pull this line low and wait until it is addressed, at this point the device will release the interrupt line. Multiple devices can have this line pulled low and the signal will only clear (return high) when all devices have been serviced.

11.1.2. Speed and Timing

The SM bus is a critical part of a computer motherboard and an additional mechanism has been implemented that can allow a device to come back out of deadlock. Where I2C allows you to lower, even stop the clock, this is not allowed on an SM bus. When the SCL line is in logic 0 state for longer than 35 mS a bus timeout occurs. The SM bus specification states that when such a condition occurs, all devices on the bus will reset and release SDA and SCL. Due to this timeout mechanism the SM bus cannot be clocked slower than 10KHz. Frequencies lower than that would cause the SCL low period to become longer than 35mS and a reset would occur.

The rise and fall time of the SDA and SCL signals is controlled more tightly.

Normally this does not pose problems; most I2C devices will happily cooperate on the system management bus.

11.1.3. Software Layer

A software layer has been implemented on top of the physical I2C protocol. This has some impact on I2C devices that are attached to the bus. Elements such as repeated start and arbitrary packet lengths are not allowed. Careful checking needs to be made to verify that an I2C device will be able to interoperate on the SM bus. Normally there is no problem, but for certain devices there may be problems.

The software layer also allows for packet error checking, as a CRC 8 checksum is appended after the last byte of the payload. This allows devices to verify the integrity of the packet.

Packets on the SM bus cannot be longer than 32 bytes.

11.1.4. Application

The SM bus is used on a computer motherboard for a number of tasks, most notably is the temperature monitoring. The LM75 I2C thermometer was specifically designed for this purpose. Besides temperature monitoring the SM bus is also used to read various voltages on the motherboard and control peripherals like audio interfaces for example.

Another key application is reading the memory configuration, as every DIMM module has an I2C EEprom onboard that contains timing information for the memory module. The motherboard retrieves this information so that it can set up the memory controller with the correct timings.

The PCI bus provides access to the SM bus signals as well. It is possible to install an EEprom containing information about the particular card (vendor id and product id) for example.

11.2. PMBUS

The power management bus is an extension of the SM bus that was first defined in the mid 2000's, while the final specification was released in 2007. It pushes the clock speed up from the 100 KHz maximum under SM bus, to 400 KHz. The packet sizes can be up to 256 bytes and there is support for 'broadcast' messages.

The multi master mechanism of I2C is fully exploited. Even where the SM bus has this capability it was almost never used, the PM bus uses this to send information between intelligent masters.

As with SM bus, PM bus also defines a software layer on top of the I2C protocols. Devices are divided into classes and each class has a number of commands. A device belonging to a certain class must implement the commands that are applicable to that class. On top of that a device may implement additional commands that are vendor specific.

Notably smart batteries are using the PM bus to inform the attached device, most commonly a laptop, about the state of the battery charge.

11.2.1. Applications

The PM bus is used to control and monitor the power in a computer or computer system; it can also be used to control the power up and power down sequence of the system. While running, the PM bus is used to monitor voltages and currents.

11.3. IPMI

The Intelligent Platform Management Interface takes it a step further. This implementation allows I2C to run throughout system chassis such as server racks. Every shelf or appliance in the rack has its own controller. These controllers are interconnected using the IPMI bus to a chassis controller.

Typically chassis controllers feature a network port (ethernet0 that allows an outside control program to issue commands. IPMI is used in server farms for so called lights out management. Simply install the server rack, plug everything in and walk away. You can control the entire rack remotely through IPMI. The Ethernet side of IPMI is called RPMI (remote management control protocol).

Using this system you can control the power of individual elements in the system. Also you can retrieve critical parameters such as power consumption and uptime, but you can also force reboots of crashed systems.

IPMI provides self diagnostic and automatic messaging when a problem develops. The Baseboard Management Controller (BMC) is the core of the system. This controller has I2C connections to satellite controllers that control one shelf or subdivision of the rack.

11.4. ATCA

The advanced Telecommunications Computing Architecture is a specification crafted in an attempt to standardize the computing infrastructure used in telecoms equipment.

Telecoms equipment such as exchanges, switches, routers and access multiplexers use a blade like construction. Each blade holds either one or more line interfaces, or a support function. Every manufacturer has his own definition of connections and hardware. This leads to numerous problems in interconnecting equipment from different vendors, not only on the electrical but also on the mechanical front. The ATCA defines a common mechanical chassis with fixed size, slot count and connecter pin out. The standard incorporates an opening to install custom connectors or backplanes to accommodate various uses. The cabling in the chassis is split between zones. One is used for power distribution, one for board to board communication and one for application specific signals. ATCA also specifies the usage of IPMI (see previous topic) to manage individual boards and features of the chassis. A shelf controller has redundant I2C busses and uses the IPMI protocol on top of the I2C bus to manage the shelf. The shelf controller itself interfaces to a control network using Ethernet.

12. ELECTRICAL SPECIFICATIONS OF THE I2C BUS

As the chips designed for an I2C bus can function on different Supply voltages the following levels have been set. The I2C specification has seen a number of revisions and the original criteria no longer apply.

PARAMETER	SYMBOL	STANDARD MODE		FAST MODE		UNIT
		MIN	MAX	MIN	MAX	
LOW level input voltage (fixed value: old spec)	Vil	−0.5	1.5	N.A.	N.A.	Volts
LOW level input voltage (VDD related: new spec)	Vil	−0.5	$0.3 \times V_{DD}$	−0.5	$0.3 \times V_{DD}$	Volts
High level input voltage (fixed value: old spec)	Vih	3.0	VDDmax	N.A.	N.A.	Volts
High level input voltage (VDD related: new spec)	Vih	$0.7 \times V_{DD}$	VDDmax	$0.7 \times V_{DD}$	VDDmax	Volts
Fall time from ViHmin to ViLmax	toff		250		250	nS
Capacitance of signal pin			10		10	pF

For other electrical specifications please refer to the component datasheets or the full I2C spec. The number of interfaces connected is limited to the number of available addresses and the load capacitance on the bus. This total bus capacitance may not be higher than 400pF. In the new standard this is preferred to be less than 200pF. .

12.1. ENHANCED I2C (FAST MODE)

Since the first I2C specification release (which dates back from 1982) a couple of improvements have been made. In 1993 the new I2C spec was released. This new specification contains some additional sections covering FAST mode and 10-bit addressing. In the following section the Fast mode will be covered, while in the section after that, information about 10-bit addressing will be given.

In the FAST mode the data rate has been increased to 400 Kbit/s. To accomplish this task a number of changes have been made to timing.

Since all CBUS activities have been cancelled, there is no compatibility anymore with CBUS timing. The development of ICs with a CBUS interface has been stopped, and the existing CBUS IC's are being taken out of production.

The inputs of the FAST mode devices all include Schmitt triggers to help to suppress noise. The output buffers include slope control of the falling edges of the SDA and SCL signals. If the power supply of a FAST mode device is switched off the BUS pins must be left floating so that they do not obstruct the bus.

The pull up resistor must be adapted, for loads up to 200 pF a resistor is sufficient, but for loads between 200pf and 400pF a current source is preferred. There are specific current source chips available for this purpose such as Linear Technologies LTC4311.

12.2. EXTENDED ADDRESSING

Due to the popularity of the I2C bus the address space has been completely exhausted. Many devices have overlapping address allocation and although therefore newer parts allow for a very wide range of address selection using various signal levels on the address bits or the usage of creative constructions by tying the address lines to other signals.

Even this does not always solve the cramped address space. Therefore the I2C standard has been adapted and a 10-bit addressing scheme put into place.

A chip that conforms to the new standard therefore receives 2 address bytes. The first address byte has its 5 MSB's set to 11110, followed by the top 2 address bits of the 10-bit 1ddress. The last bit is as usual the R/W bit.

All devices that understand this mechanism will respond with an ACKNOWLEDGE. Other devices that do not understand 10-bit transfers are now automatically out of the loop, since these do not have an address that starts with 11110.

The second byte sent contains the remaining 8th bit of the 10-bit address. At this point only one device will send acknowledge, and operation continues normally.

This mechanism is totally transparent for I2C devices and no special measures need to be taken when having both 8 and 10-bit addressable devices on the bus together.

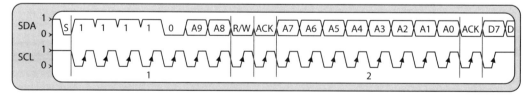

The I2C system actually allocates the combination 1111x for the 5 MSB's as reserved for 10-bit addressing, but actually only 11110 is used for this purpose. This was changed through revisions of the standard simply because of the fast address allocation that came with its growing popularity. The address combinations starting with 11111 are therefore now reserved for future enhancements.

13. DESIGNING THE BUS SYSTEM

When designing the I2C section of your project there are a number of things to consider besides the address space layout, the bus speed and topology you must also consider other key factors like the dimensioning of the pull up resistors, single or multiple busses and mixed voltage operation.

13.1. ADRESSING

The first thing that comes into mind when selecting I2C capable integrated circuits is the address mapping. Will I be able to fit all the functions into the allowed address space? Fortunately I2C devices are somewhat categorized according to functionality. There is a large amount of certainty that your memory devices are not going to occupy the same address range as your I/O expanders. A sheet of paper helps you to build the address map.

It gets a bit tricky within a group. Mixing different parts that fall in the same category can be problematic and the possible addresses can be exhausted quickly. Some devices come with alternate addresses that are indicated by a suffix to the part number (for example the PCF8574 and PCF8574A come to mind. These devices are identical apart from their base address.).

When the available address range is exhausted you will have to resort to a different approach, you can either create two or more independent busses or you can multiplex them.

13.2. BUS EXPANSION

Depending on how many devices are going to be attached, and what their relative distance is from the master, it may be required to think about bus buffering and or expansion. Bus expansion is covered in another section in this book and there are a number of devices readily available to do this.

If the I2C bus needs to go off board then the cabling becomes an issue as there is only so much the extender and expander chips can do. The cabling itself, connectors and cable type, become a point of further attention.

The next chapter goes into detail on the common problems you are faced with in creating the bus system.

14. F.A.Q.: HARDWARE

This section will give answers to a number of common questions that reoccur frequently when dealing with the I2C bus.

14.1. WHAT IS THE MAXIMUM ALLOWED LENGTH OF THE BUS?

This depends on the load of the bus and the speed you run it at; in typical applications you can easily span a few feet (2 to 3 feet or up to 1 meter). A better definition would be: the maximum distance is determined by the maximum capacitive load that has been specified (see the electrical specs earlier in this book).

If you run at a lower clock frequency you can typically cross a longer distance. If you are careful in routing your PCB's and cabling them, you can take it pretty far. I have seen systems where I2C travels over several tens of feet (6 meters +) reliably. If you take care of the PCB layout and keep SDA and SCL away from power circuitry and terminate them correctly with an appropriate pull up resistor, you can go quite far.

Once you leave the PCB you can use a signal ground approach. If you can use a twisted pair cable it gets even better. You can twist SCL with GND and SDA with VCC. That way you can easily carry I2C signals inside a system like a telecoms or computer rack (19 or 24 inch rack systems). I know several industrial machines that use I2C as internal communications protocol and have tens of feet of cabling inside.

A cheap solution is to use standard Ethernet cable and jacks. These kinds of cables have 4 twisted pairs and by employing a smart connection mechanism this can be exploited to also transfer power and ground to remote nodes.

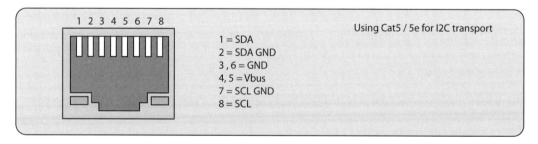

If you need to go far at high speed then you can use an active current source. NXP and others have several products for this purpose. There are even devices that allow the shifting of I2C signals to larger voltage levels to bridge long distances. If ground loops are an issue you can actually make a galvanic isolation for an I2C bus. More on that subject later.

The last thing to be taken care of is the bus termination. Long lines will cause reflections. These reflections can cause 'ghost signals' to appear. This can be overcome by using a charge pump mechanism like the high speed devices have. It may be beneficial to insert a small series termination resistor on the bus lines as well.

14.2. I WANT TO TRANSPORT I2C OVER A LONG DISTANCE OR OFF BOARD

There are multiple solutions to that. Some people use off the shelf USB or Ethernet cable to accomplish this. NXP has special buffers like the P82B96 and 82B715 to send USB signals over long distance.

If you need to go off board then signal integrity becomes an issue. Not only should the cable be shielded, the SDA and SCL lines should be kept apart from each other to avoid possible crosstalk. It is recommended to use a twisted pair type cable and twist SDA with ground and SCL with power. If you use 8 conductor cables you can reserve one twisted pair for power, one for ground and individually pair SDA and SCL with their own signal ground.

If you want to use USB cables you need to send SCL over the power wire (vBus) and SDA over either D+ or D. Tie the other two wires to ground.

For on board, you can use an active pull up device like Linear Technologies LTC4311 or you can build one your own with the following schematic:

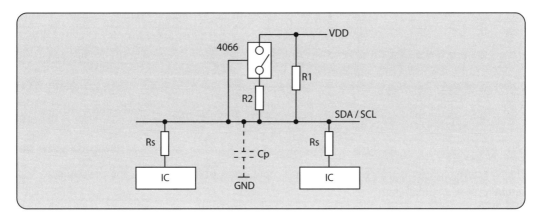

Now how does it work?

First of all Rs: this is a series resistor used to minimize cross talk and undershoot. It also protects the I/O drivers of the I2C devices against over voltages and over current effects. These resistors are advised if you run a long bus at high speed (such as in enhanced I2C mode). When the bus becomes free all output stages on the bus are turned off and SCL (or SDA for that matter) become high.

This will not happen immediately, but the voltage will rise during a specific short time. Now suppose the switch is not there. The charge time of the bus capacitor is determined then by the value of R1 only. The higher the resistance value is, the longer it will take for the bus to reach a sufficiently stable High. We can't make the Series resistor too small because we will then go out of spec of the maximum allowable current into one I2C device when it turn's its output driver on.

When we calculate for a current of 3ma we end up at approx 1800 ohms for the series resistance. 5volts / 3ma = 1666 Ohms.

To stay somewhere below this 3mA rating we should take 1800 Ohms. The charge time for a bus capacitance of about 200pf would be around 360 nS. That is out of spec. The Spec for the rise and fall time in Fast I2C is set to approx 300nS.

But we can't drop the value of our resistor without breaking the other spec. of 3mA Max current.

Now if we had a means to change the value of the resistor temporarily. . . . and that's exactly what is done with the Analog Switch. If the voltage level sensed by the switch is in the range 0.8 to 2 volts then it will turn on. Meaning that the moment the voltage on the SDA line starts rising, this resistor will kick in, reducing the resistance and increasing the charge current for a 5volts supply/ (1K8//1K2) = 720 Ohms. The charge current will then rise to 5 volts / 720 Ohms = 7 mA. This is allowable for a brief period of time. Of course all of this is a dynamic process and the actual charge current will change due to the fact that the bus voltage will rise.

A small graphical representation will explain more:

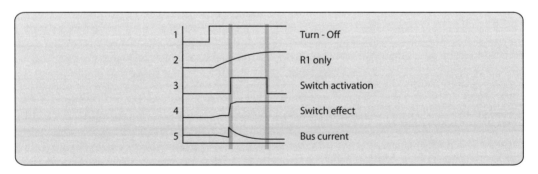

The first trace represents the turning off of the output driver of the I2C device and the bus is returning to a logic high state.

The second trace is what you get if you only use a resistor. The bus slowly comes up to 5 volts due to the RC constant of Rp and Cp. This parasitic R C time constant slows down the rising edge.

Trace number 3 shows the kicking in of the Analog switch. If the bus gets to approx 1 volt, the switch kicks in and switches R2 in parallel with R1 thus lowering the overall resistance and beefing up the current. The switch opens when the bus reaches approx 3 volts and removes R2.

Trace 4 shows how the voltage on the bus changes. You can clearly see that it rises much faster when the switch is turned on.

The last trace shows the current flowing into the I2C device. It starts at approx 3mA.

When the output stage is turned off, the current drops slightly due to the fact that the voltage on the bus is rising, but the moment our switch kick's in you see the current double. The same as before the switch closed: the current drops as the bus voltage rises. When the switch opens again the current drops a little to charge the capacitor up to 5 volts, but by that time all the chips have already detected a logic one and we are still well within the 300nS rise time window.

It is important not to treat this 4066 as a digital switch when looking at this circuit. We are only using one of the internal MOSFETs in the device. This circuit works because of the internal construction of the 4066.

14.3. I WANT TO EXTEND IT "BY THE BOOK". IS THERE SOMETHING LIKE A BUFFER FOR I2C?

Yes this does exist. NXP and several other manufacturers have several devices to buffer the bidirectional lines of the I2C bus. Typically this is a current amplifier. What it does is to force current into the wiring (a couple of mA). That way you can overcome the capacitance effects of long wiring.

Let's look at a couple of devices of interest.

14.3.1. NXP P87B715

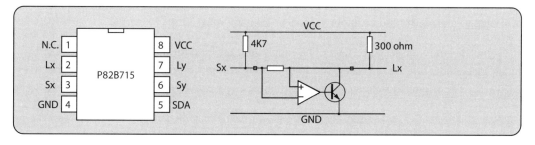

The P87B715 is a buffer that allows the bi directional I2C signals to pass. It is essentially an impedance transformer. The load capacitance is roughly divided by 10, so a bus that is loaded with 3000pF acts the same as a bus loaded with only 300pF.

The S x and S y lines are the normal I2C lines and the Lx and Ly are the current amplified lines

This device essentially senses the current that is being pulled and amplifies it when necessary. When the S x line (the non amplified side) is being pulled low there will be a current through the sensing resistor. The opamp will detect this current (the current flows from VCC, through the 300 ohm sense resistor into the pin driving the S x or S y signal) and turns on the transistor. This will now pull the Lx or Ly line low. When the signal S x is released the current stops flowing and the opamp turns off the transistor to stop it conducting.

Since the device is an open collector the high level will be determined by whatever voltage you apply on the pull up resistors outside the device. The power supply to the devices is only used to power the sense opamp. A long as this supply is higher than the highest voltage this system will work. The P82B715 operates from 3 to 12 volts.

However, you will need this component on both sides of the line. The charge pump in these devices can deliver currents up to 30mA. And that is way too much for a normal I2C chip to handle. This device has a wide operating voltage. This allows the low impedance side to be driven from higher voltages than the other side.

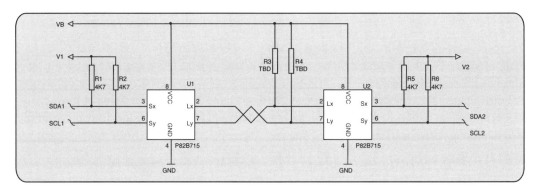

The schematic above shows the circuitry. A long cable is running between two P82B715's. These signals have their own pull up resistors (R3 and R4). The I2C bus SDA1 and SCL1 is pulled up to its own supply voltage. This can be either 3.3 or 5 or even 2.5 volts.

The same applies to the second I2C bus. The supply voltages for both sides of the interface can be totally independent from each other. The supply voltage for the interconnection between the P82B715's is also completely independent of either v1 or v2.

Suppose you have a 3.3 volt system that needs to talk over a long distance to a 5 volt system. You want to transport 9 volts across so that you can power the remote side through the cable.

In this case you would set v1 to 3.3, v2 to 5 volts and vb to 9 volts. As long as the P82b715 is powered from VB everything is perfectly OK.

The resistors R3 and R4 need to be tuned to function with the maximum current that can be pulled by the P82b715 which is 30mA. The drop over the transistor is typically around 0.4 volts. The equation is simple: $R_{pu} = (V_b - 0.4)/ 30mA$.

It is perfectly safe to connect multiple P82B715's together to form a network.

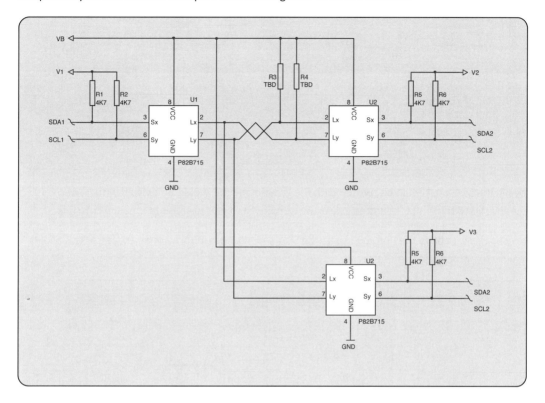

14.3.2. NXP P82B96

The P82B96 is a bidirectional I2C buffer that level shifts to high voltages. This unique capability allows transmitting I2C signals over long distances, often in excess of 20 meters (66 feet) at speeds of up to 400 KHz.

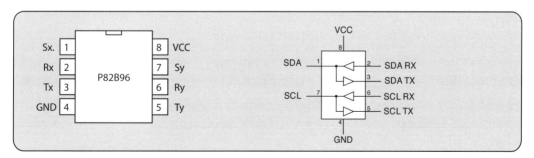

Just like with the P82B715 there are always two of these chips needed to cross the long cable. Both sides need a P82B96. The signal levels on the high current side are not compatible with the normal I2C levels and need to be dropped down again.

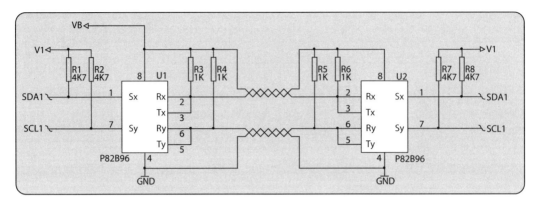

Ideally the SDA line should be twisted with the ground and SCL would be twisted with power. It is not advised to transmit SDA and SCL over the same twisted pair because of possible crosstalk.

Besides this capability the PB82B96 splits the SDA and SCL lines in forward and reverse. This allows the use of optocouplers in the circuit.

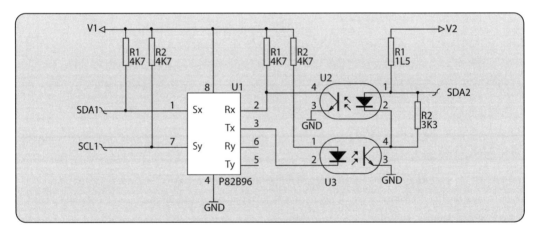

In the above image only one channel (SDA) is shown, but an identical construction with the optocouplers also needs to be implemented for the SCL signal.

14.3.3. NXP PCA9600

This is essentially an enhanced version of the P82B96 that can drive even higher capacitive loads and can work up to 1MHz (fast mode+). Using the right cable and an appropriate clock speed you can bridge several kilometres with these chips. Just like with the PB82B96 you need one of these transceivers on both sides of the cable. The supply voltage of the bus can be up to 15 volts.

14.3.4. Hendon 5501 / 5502

These devices not only allow for the level shifting of the bus but also provide a true buffering capability. Whereas other devices need a driver on both sides of the connection, the Hendon 5501 and 5502 do not require a counterpart on the other side of the connection. They are effectively bus buffers.

The devices are capable of level shifting the I2C signals even above the devices own supply voltage. An enable pin allows for the controlled shutdown of the buffer. This allows you to perform multiplexing on the bus.

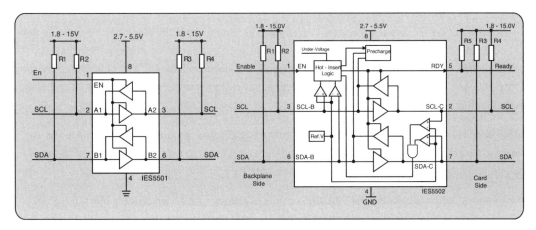

An advantage of the IES5502 is that it allows for hot plugging devices on the bus. When the slave side is removed it is guaranteed not to lock up the master side of the bus. It also allows the master to signal to a slave device that the system is ready for power up. These devices allow level shifting for up to 15 volt signals.

14.4. CAN I ISOLATE AN I2C BUS? (USING OPTOCOUPLER OR SOME OTHER MEANS)

This is possible. Besides the previously mentioned PB82B96 and PCA9600 and some optocouplers, there are also some specific chips that fulfil this functionality. You can of course brew your own circuitry as well but let's take a look at the off the shelf circuitry first.

14.4.1. ADUM1250 / ADUM1251 / ADUM2250 / ADUM2251

These devices have a built in galvanic isolation that uses chip scale transformer technology. The devices provide an effective isolation barrier of 2500 Vrms for up to 1 minute (according to UL1577). The circuits do not need external components to provide the signal bridge, everything is integrated, and the chips feature a split power supply and work between 3 and 5 volt supply.

The ADUM1250 offers two fully bi directional channels whereas the ADUM1251 has only one bidirectional and one unidirectional signal. The unidirectional signal can be used for SCL in systems that do not need multi master or clock synchronization mechanisms.

The ADUM 2250 and 2251 are upgraded versions that offer up to 5KV of isolation. The body is wider to allow for enough physical distance between the two different domains.

More information will be given later on in this book.

14.4.2. Homebrew

The circuit is rather complex due to the bidirectional nature of the I2C BUS. Actually there are a number of solutions here.

There are plenty of circuits out there that attempt to provide a home brew solution to isolating the I2C bus. Most, if not all of these circuits are plagued by glitches that occur when changing levels. These glitches occur due to the propagation delays in the isolator elements. The bidirectional nature of the SDA and SCL line results in an outgoing signal having to come back across the isolation barrier as well.

Other circuits have the problem that they can go into deadlock or produce plainly wrong levels on their outputs. The problem always sits in finding the correct way to split the bidirectional signals into two unidirectional lines, and then preventing deadlock from happening due to feedback.

The easiest solution is by using the PB82B96 or PCA9600 devices and two optocouplers. If higher speeds are required you may need to grab a hold of active optocouplers like the 6N137. Driving the led in the optocoupler may be troublesome as well. At one point there were active optocouplers like the 74OL6000 but they seem to have gone the way of the wind.

The base schematic has been given before in the topic on bus extension. The drawback of that circuit is the relative slow clock speeds due to the driving of the LED.

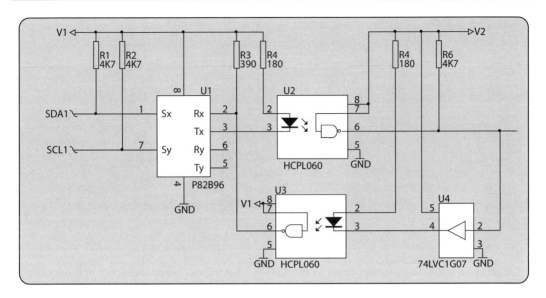

The above schematic solves the driving problem.

14.5. WHAT IF I RUN OUT OF ADDRESSES? IS THERE AN I2C ROUTER?

There is no such thing as a router in the literal sense. However, it is possible to switch the I2C bus signals around and there are multiple ways of doing it.

When you run out of available address space, for example you need to control 16 or more PCF8574; you can spread the devices between multiple busses. By selecting which bus, and thus which group of devices, you can create virtual sub addresses. Think of it in terms of bus x, device y.

14.5.1. Using Generic Multiplexers

Of course you can use special devices such as the PCA951x family, but it can also be done in a different way. Essentially any device that can switch a bidirectional signal will do. The question is where do you find such a thing? Well, look no further, any device made to switch analog signals under digital control will do. The 4051, 4052, 4053 and 4066 come to mind. These CMOS analog switches are essentially a MOSFET where the drain and source are brought out. Under control of a digital signal you can bring the MOSFET to either conduct or isolate. The polarity or direction of signals doesn't matter. The ideal device is the 4052 which is a 1 to 4 multiplexer. You can also use the DG409 dual 1 to 4 analog switch.

The 4052 is also available as a 74xx4052 to accommodate low voltage operation. Devices such as the 74LVX5052 can run on voltages as low as 2.5 volts. The 74VHC4052 can even drop to 2 volts if needed.

By connecting the master I2C pins to the common nodes of the multiplexer you can create 4 independent I2C busses. Selection is done by simply applying 2-bits to the select inputs of the 4052.

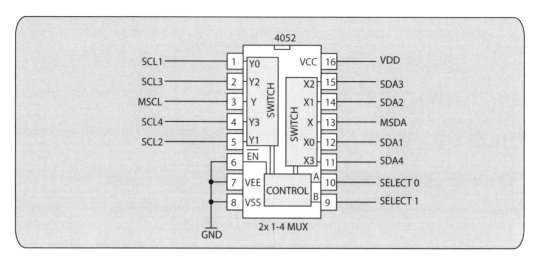

Of course every I2C channel must have its own pull up resistors. The series resistance of the analog switch is in the order of 50 ohms and therefore negligible for our applications.

14.5.2. Using Dedicated I2C Multiplexers

NXP and others have several I2C bus multiplexers that offer capabilities far beyond that of a simple analog multiplexer. The PCA9516 / PCA9518 and PCA9544 are built specifically for the purpose of multiplexing the bus. Some of these devices also allow for the multiplexing of an interrupt signal. Keep in mind that these devices, just like the solution of using an analog multiplexer, does not buffer the I2C bus. Whatever capacitive load is present will be replicated.

14.6. LEVEL SHIFTING THE I2C BUS

In some applications there may be different voltage domains and it will be necessary to create the appropriate voltage levels. Level shifting becomes a problem due to the bidirectional nature of the signals. Fortunately NXP and others also have an answer to this problem as well in the form of the P82B715.

It is also possible to simply make a level shifter with a couple of simple MOS transistors.

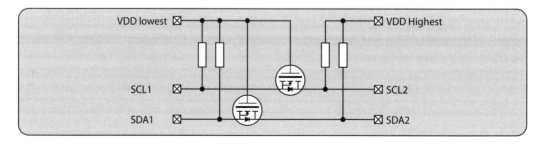

How does it work? If nothing is active on the bus, that is nobody is pulling a line low, the SDA and SCL signals on both sides are pulled up by their resistors to the correct levels for that side. The N channel MOSFETs are off since their Source voltage is roughly the same as the gate voltage. When a signal from the left side goes logic low the MOSFET connected to it turns on, thus pulling its drain (in this case the right hand side) also low.

When a signal from the right hand side pulls low this low is passed through the body diode of the MOSFET.

This system only passes logic LOW levels and it is the job of the pull up resistors to create the high levels.

As long as the right hand side (the side connected to the Drains of the MOSFETs) runs at an equal or higher voltage than the left hand side this system will operate. A requirement is to have a MOSFET that has a Vt roughly 1 volt below the lowest VDD. If you run it at 3.3 volts on the low side, the Vt of the MOSFET (also called the Threshold voltage, or the voltage required to turn the MOSFET on) should be around 2 volts.

Almost any small signal MOSFET such as a 2N7002 or BSS138 will do fine.

This circuit has one problem though. If either side is powered off, this will lock up the bus on the other side. To prevent this you can insert 2 more MOSFETS.

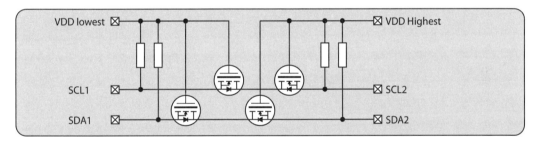

14.7. IS THERE SOMETHING LIKE A STANDALONE I2C CONTROLLER?

Suppose I don't want to emulate the bus by software (bit banging) or I don't have an I2C interface on my processor. How do I interface to the bus?

There are special devices that allow you to perform exactly this functionality. The PCF8584 and PCA9564 incorporate a complete I2C interface. These chips have a standard 8-bit data bus and some control signals. Entering into full details on the operation of these devices would require half a book by itself, but NXP has some excellent application notes as well as sample code on how to use these devices with a range of microcontrollers. A word of caution though, the PCF8584 was designed before the full I2C bus protocol was finalized and is notorious for the bugs it contains. NXP has a special application note that outlines all the revisions of this chip and all the behavioural changes in it.

14.8. WHY DOES THE CLOCK LINE NEED TO BE BIDIRECTIONAL?

The clock line needs to be directional when using a MULTIMASTER protocol and when using the synchronization provisions of the I2C protocol. In a single Master system it is not required to have a bidirectional SCL line, since in this case the clock will always be generated by the Master and as you only have one of those on the BUS there is no problem.

If you run in multi-master mode then this changes, each Master must be able to receive data from the other master, and at the same time it must be able to check the Clock line too. For more information about bus synchronization check out the topic dedicated to it.

15. F.A.Q.: PROTOCOL

This section will cover a number of common problems that arise when dealing with the protocol.

15.1. Q. HOW DO I GENERATE A REPEATED START CONDITION AFTER THE LAST BYTE?

When making the SCL high my device pulls SDA low to acknowledge. So far no problem but how do I make a new start now? The device is pulling SDA low!

A. Assuming that you are the master and have just sent the last byte you are now in the acknowledge cycle (you have sent the last byte since the slave is pulling SDA low during acknowledge).

The solution is simple: you always have to complete your ACK cycle (6). So, simply make the SCL low. The slave device will now release SDA at this point. Now you can issue a repeated START event (7) and you do this by raising your CLK line, waiting the required minimum time and then lowering the SDA, waiting again for the required time and finally lowering your SCL. This will create the repeated start.

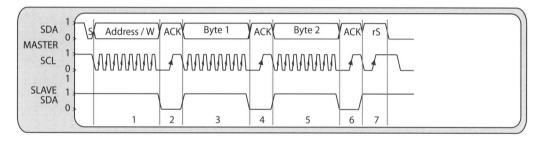

In short: a repeat start can only be created after a valid ACKNOWLEDGE cycle.

15.2. Q. WHAT IF I WAS IN RECEIVE MODE, HOW DO I CREATE THE REPEAT START?

A. When you were on the receiving end and have just read in the last bit you should normally give an ACKNOWLEDGE to the slave talker. To issue this acknowledge event you would lower SDA and give a clock pulse. But the slave will begin transmitting the next bit and this may prevent you from sending a repeated START event.

In this case you approach this from a different angle. Instead of lowering SDA as with a normal give ACKNOWLEDGE (4) you leave SDA high (6). Now, raise the SCL line, wait for the required minimum time and then lower the SDA. Wait again for the required minimum time and the lower SCL. That's all. Essentially you are sending a nACK telling the slave that you don't need any more data (nACK is sent by leaving SDA high while sending the clock pulse). However, in the middle of the nACK cycle you create a repeat-start by dropping the SDA while the SCL is high (6).

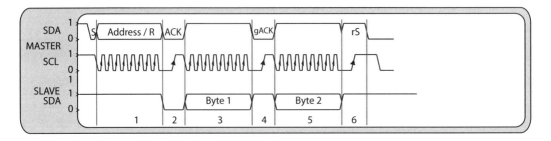

15.3. Q. IS IT OKAY TO ABORT AN ONGOING TRANSMISSION ANY TIME?

A. According to the specification this should work, but it depends on the IC. A really I2C compatible IC will be able to handle this. Whenever a START or STOP condition is detected, the internal logic state-machine of the device is forced into a certain state. Internally the part that detects START and STOP is different from the logic that does all other processing. The START together with the address register is to be considered as a functional unit inside the chip. So is the STOP detection circuitry.

When a START is detected all internal operations are cancelled and the chip will compare the incoming data with its own address. When a STOP is detected ALL chips on the bus will reset their internal logic to IDLE mode. This is also used to cut power consumption. When a STOP is detected all logic is shut down except for the START detector. When a start is issued on the bus the START detector will 'wake-up' the rest of the internal logic. The STOP event may also be used to trigger other internal systems in the device.

There are devices out there that use the terminology 'two-wire bus' or some other description to avoid licensing the standard, but these devices may or may not be fully I2C compliant. Also software implemented slave devices may not have the appropriate flow control to be able to handle such conditions. Check it out in your application by simply trying this.

15.4. Q. DO I NEED TO GIVE THE ACK IN READ MODE ON THE LAST BYTE?

My chip starts sending data and occupies the bus… how do I send a stop now?

A. After you have read the last byte from the slave you must send a nACK condition. This is similar to an ACK with the exception that you leave the SDA line high during this cycle. This signals the slave device that you are not ready. You can now create a stop.

16. Q&A SECTION: TROUBLESHOOTING

This section deals with practical problems when using I2C in a system.

16.1. Q. CAN I MONITOR AN I2C BUS IN SOME WAY?

A. This is possible. There are a number of commercial I2C monitor / debuggers out there. Though these units are powerful they can be pricey.

- Corelis Buspro-I www.corelis.com
- Jupiter Instruments I2C bus monitor www.jupiteri.com
- Avit Research www.avitresearch.co.uk
- Micro Computer Control corporation www.mcc us.com
- TotalPhase Aardvark and Beagle www.totalphase.com
- Calibre www.calibreuk.com
- Joint technology www.jointechnologyhk.com

There is another possibility to do this by using the PCF8584 chip from NXP. This universal CPU to I2C interface has a special monitoring mode where it 'sniffs' the bus. You can simply read all information that flies back and forth on the bus. Using some software routines and a small microcontroller you could make a universal I2C data logger.

If you have access to a storage or digital oscilloscope you can make a simple circuit that will trigger the scope whenever a start condition is detected. If your oscilloscope does not have memory you can still get away with this trigger circuit if you put the master in an endless loop around the I2C call you are performing. In the section on debugging tools later on in this book I will explain these circuits.

16.2. USBEE

The USBee is a small logic analyzer / generator module made by CWAV based in Temecula California. http://www.usbee.com/index.html

This little unit is essentially a Cypress Ez-USB chip with some clever firmware installed that, in combination with a PC, offers a virtually unlimited memory buffer to capture or generate digital signals.

The software has special features to capture and analyze various digital formats, amongst which is I2C. The software is capable, not only of showing the waveform, but decoding the digital stream as well. The end result is a human readable representation that shows the content of the transaction.

The USBee has a virtually unlimited data buffer as your computer is used as memory space. Data is captured and streamed on the fly over a USB2.0 connection into the memory of your computer. The visualizer post processes it and displays the received signals.

Dedicated processors are available to decode various serial protocols on the fly. The above image shows a snapshot of two I2C transactions occurring back to back. The system has recorded the digital patterns for SDA and SCL on the first two lines. The third line shows the decoding of the transaction independent states for START, STOP, ACKNOWLEDGE. Address and data operations are recognized and shown in time step with the captured data.

Besides I2C the device also handles a wide range of other common protocols such as SPI, Microwire, Can, RS232, USB (low speed only), I2S, SMbus , Ps/2 and you can even write your own interpreter.

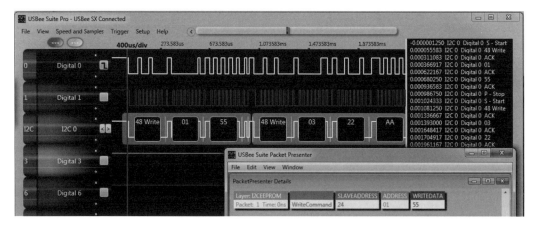

There is a 'Pro' version available that adds multiple visualization options as well as a packet presenter. This software module allows an even deeper analysis of the data stream. The default software can show you base events and raw hex data, but the packet presenter can analyze this and present the result in plain English text. It is possible to show symbolic names for register numbers and data fields in the serial stream. Instead of reading 'Write 3, 29' the packet presenter may show something like: Device ADC: Write Command: Control register, Channel 9, Start Conversion.

The packet presenter uses a template where you describe the device in terms of registers and data fields. You can then apply this template to the captured data and the presenter will visualize detailed information. This is a very useful system when debugging large systems with multiple devices as you don't need to decode by hand all bytes flying back and forth.

16.3. Q. IS THERE A WAY TO TEST/DEBUG I2C BUSSES?

A. There is no pre-packaged way to debug an I2C bus; however a simple flowchart may help to get it running. I am assuming that there is only one master connected in the system, if there are multiple masters disconnect all but one.

16.3.1. Hardware Check

1) Make sure the bus is not 'stuck' low. This could be the result of a bad power supply (chips go into latch up during power-on) or a bad chip. Both SDA and SCL should be logic high when there is no activity. The high voltage on the bus must be higher than the minimum required of any of the attached slaves.
2) Check for the presence of pull-up resistors of an appropriate value and make sure they are connected to the highest bus voltage in the system.
3) Verify that you can bring SDA and SCL logic low. You can easily do this with a piece of wire by shorting one line at a time to ground. Be on the lookout for possible shorts between SDA and SCL. Keep an eye on the power consumption of your system. If that jumps up by more than a few milliamperes you may have a push-pull driver on your bus. Both SDA and SCL must be open-drain or open-collector outputs. You are NOT allowed to actively drive the bus high. A lot of master implementations in microcontrollers that don't have hardware I2C use push-pull drivers for SCL and SDA. It is prudent to disconnect the system master before doing this test.
4) Have the master bring SDA and SCL low and verify the low level. If this is above 0.5 volt then the pull-up resistors are too low a value and you are injecting too much current into the output drivers. Increase the value of the pull-up resistor to lower the pull-up current.
5) If you have access to a scope: hook it up and have the master toggle SDA and SCL lines repeatedly. Verify the rise-time of the SDA and SCL signals. You may need to beef up the pull-up current by lowering the value of the pull-up resistors. The minimum allowable resistor for a 5 volt drive I2C bus is 5volts / 3mA = 1600 Ohms. A typical value of 4700 ohms should do fine.

16.3.2. Communication Check

At this point we have an electrically correct bus with no shorts. Now it's time to see if we can send something over the bus. Again I am assuming that there is only one master connected in the system. If there are multiple masters disconnect all but one.

1) Attach a known good slave device like a PCF8574 with some LED's attached and perform a write.
2) If you have a scope: capture the data flying by and verify that you get an acknowledge.
3) If no acknowledge is received, verify the address you are accessing: 7 bits followed by the Read/Write bit. Double check the device address that was given and remember the potential confusion of a 7 bit address in an 8-bit byte. You may have to shift the address left by one bit.
4) Check that the address selection pins of the slave device are tied correctly and that you are transmitting on this address. Try alternate addresses.
5) If multiple slaves are connected: disconnect all but one and try communication again. One slave may be blocking the bus halfway through the communication.

16.3.3. Corrupted Transport

Again, if there are multiple masters disconnect all but one.

1) It is possible you have two devices with the same address. Double check the address allocation.
2) Disconnect all devices and then add them one by one, testing communication at each step. The step where communication fails either occupies an address that is already in use or is simply a defective part.
3) If communication is only corrupted once in a while, you may have noise on the SDA or SCL line. It is also possible that the bus load is too high and that the rise time is impaired. Hook up a scope and verify this. Either adjust the value of the pull-up resistors or add an active pull-up device.

16.3.4. Multimaster Trouble

An I2C bus with more than one master can be very tricky to debug. There is a lot of generic I2C code out there that does not fully implement the multi-master mechanisms correctly.

1) Check slave communication one master at a time. Remove the non talking master(s) physically from the bus. Once both, or all, master firmware codes have been verified independently you can proceed. Verify that each master, by itself, is capable of performing correct communication with the slaves it needs access to.
2) Verify that your master can correctly detect and handle an external device that pulls SDA low while you are sending logic 1 over the bus. Your master code should stop transmitting, tri-state both SDA and SCL and wait for a STOP condition to occur on the bus. Put your master in a transmission endless loop and simply tap a ground lead on the SDA line. The master should stop talking and both SDA and SCL should return to logic high.

16.4. Q. I WANT TO EXPERIMENT WITH I2C, ARE THERE DEMO KITS AVAILABLE?

In the section on bus monitors there is plenty of information on such devices. The cheapest ones can be had for a few tens of dollars and allow I2C bus exercising from a PC platform (most commonly USB to I2C these days although there are also PCI cards available).

NXP and other manufacturers have some demo kits that go beyond a simple master. The NXP kits feature a number of slave devices that allow you to speed up code development.

Kits such as the NXP OM6275,598 (Digikey 568 3615-ND) can be had for under 50$US at the time of writing and features a range of devices such as the PCF8574TS, PCA9536D, PCA9540BD, PCF85116, PCF8563TD, PCA9538D, PCA9551D, SA56004ED, PCA9543AD, PCA9531D, and PCA9541D/01.

If you want to control I2C from your pc then the USB adapters are the easiest to use. If your PC still has a printer port then you can even use that. Sample code and a schematic will be given later in this book.

17. AN I2C DRIVER IN PSEUDOCODE

This section covers a sample I2C driver. It is written in Pseudo Code which is an imaginary programming language that any programmer should be capable of porting to his/her favourite language.

First we will define a set of basic interface routines. All text between each / is considered as remark. The following variables are used:

n, x = a general purpose BYTE

SIZE = a byte holding the maximum number of transferred data bytes at a time

DATA (SIZE) = an array holding up to SIZE number of bytes. This will contain the data we want to transmit and will store the received data.

BUFFER = a byte value holding the 'just" received or transmitted data.

Driver functions

- I2C_init: initializes I2C bus
- Start: Sends a Start condition
- Stop: Sends a stop condition
- PutByte: Sends a Byte
- GetByte: reads a Byte
- GiveAck: Gives ACK to slave
- GetAck: Gets ACK from slave

High-level functions

- Read: Reads a byte from an address
- Write: Writes a byte to an address
- RandomRead: Reads a block of data
- RandomWrite: Writes a block of data

```
/$$$$$$$$$$$$$$$$$$$$$$$$$/
/****  I2C Driver      ****/
/$$$$$$$$$$$$$$$$$$$$$$$$$/

DECLARE N,SIZE,BUFFER,X Byte
DECLARE DATA(size) Array

SUBroutine I2C_INIT
        SDA=0
        SCL=0
        FOR n = 0 to 7
                SCL=1
                SCL=0
        NEXT n
        CALL STOP
ENDsub

SUBroutine START
        SCL=1
        SDA=1
        SDA=0
        SCL=0
        SDA=1
ENDsub

SUBroutine STOP
        SDA=0
        SCL=1
        SDA=1
ENDsub

SUBroutine PUTBYTE(BUFFER)
        FOR n = 7 TO 0
                SDA= BIT(n) of BUFFER
                SCL=1
                SCL=0
        NEXT n
        SDA=1
ENDsub

SUBroutine GETBYTE
        SDA=1
        FOR n = 7 to 0
                SCL=1
                BIT(n) OF BUFFER = SDA
                SCL=0
        NEXT n
        SDA=1
ENDsub

SUBroutine GIVEACK
        SDA=0
        SCL=1
        SCL=0
        SDA=1
ENDsub

SUBroutine GETACK
        SDA=1
        SCL=1
        WAITFOR SDA=0
```

```
        SCL=0
ENDSUB

/ higher level /

SUBroutine READ (Device_addr,Number_of_bytes,DATA())
        Device_addr=  Device_addr OR (0000.0001)
        START
        PUTBYTE(Device_addr)
        GETACK
        FOR x= (0 to Number_of_bytes)
              GETBYTE
              DATA(x)=BUFFER
              IF X< Number_of_bytes THEN GIVEACK
        NEXT x
        STOP
ENDsub

SUBroutine WRITE  (Device_addr,Number_of_bytes,DATA())
        Device_addr=  Device_addr AND (1111.1110)
        START
        PUTBYTE(Device_addr)
        GETACK
        FOR x= 0 to Number_of_bytes
              PUTBYTE (DATA(x))
              GETACK
        NEXT x
        STOP
ENDsub

SUBroutine RANDOMREAD  (Device_addr, Start_addr, Number_of_bytes,DATA())
        Device_addr=  Device_addr AND (1111.1110)
        START
        PUTBYTE(Device_addr)
        GETACK
        PUTBYTE(Start_addr)
        GETACK
        START
        Device_addr=  Device_addr OR (0000.0001)
        PUTBYTE(Device_addr)
        GETACK
        FOR x= 0 to Number_of_bytes
              GETBYTE
              DATA(x)=BUFFER
              IF X< Number_of_bytes THEN GIVEACK
        NEXT x
        STOP
ENDsub

SUBroutine RANDOMWRITE  (Device_addr, Start_addr, Number_of_bytes,DATA())
        Device_addr= Device_addr AND (1111.1110)
        START
        PUTBYTE(Device_addr)
        GETACK
        PUTBYTE(Start_addr)
        GETACK
        FOR x= 0 to Number_of_bytes
              PUTBYTE (DATA(x))
              GETACK
        NEXT x
        STOP
ENDsub
```

/ $$$$$$$$$$$$$$$$$$$$$$$$ /

Some notes about the high level routines. The READ and WRITE routine read or write one or more byte(s) from/to a slave device. Generally this will be used only with 'number_of_bytes' set to 1. An example:

```
PCF8574=(0100.0000)b
CALL READ(PCF8574,1)
result = DATA(0)
```

This will read the status of the 8-bit input port of a PCF8574.

```
DATA(0)=(0110.01010)b
CALL WRITE(PCD8574,1,DATA())
```

This will write 0110.0101 to the 8-bit port of the PCF8574.

When do you need a multi-read? Consider a PCF8582 EEPROM. You want to read its contents with just one READ command.

```
PCF8582=(1010.0000)b
CALL READ(PCF8582,255,DATA())
```

You can do the same with a WRITE for the EEprom with the restriction that the 'number_of_bytes' is not larger than a page size and does not cross a page boundary. For this you will have to check the components datasheets.

The most useful instructions are RANDOMREAD and RANDOMWRITE.

Write 4 bytes of data to location 20h of the EEPROM

```
DATA(0)=(1010.0011)b
DATA(1)=(1110.0000)b
DATA(2)=(0000.1100)b
DATA(3)=(1111.0000)b
CALL RANDOMWRITE (PCF8582, 20hex,3,DATA())
```

The same goes for reading 16 bytes from the EEprom starting at address 42h

`CALL RANDOMREAD(PCF8582,42hex,15,DATA())`

The results are stored in DATA. All you have to do is read them out of the array to process them. When you give the device's address to these routines you don't have to worry about the R/W flag, it will be automatically set to the right state inside the routines.

18. DEBUGGING TOOLS

In this section I will shed some light on a couple of simple debugging techniques and tricks that will help you diagnose trouble on a live I2C bus. Of course if you have access to a storage scope or logic analyzer this makes life a lot simpler. Unfortunately, even in this day and age, not everyone has access to such machinery, and even if you do then these tricks will still help speed up the debug process.

So, you have an I2C bus and you want to monitor its activity and using a regular oscilloscope is rather tough. I2C communications tend to be of a volatile nature so essentially some form of storage is a must.

If you are in control of the master, for example because you write the master code, then simply put the failing command in a loop. By toggling an I/O pin on the processor you can trigger the scope. Put the scope in triggered mode and simply repeat the same command over and over. If your refresh is fast enough you will get a stable image from which you may learn something.

In cases where you cannot control the master code, or you don't have a free I/O port, the addition of a little trigger circuit may help.

18.1. I2C TRIGGER GENERATOR

This simple circuit built around a 7473 or 74112 JK flip-flop allows any scope to trigger on an I2C transmission. Each time a start condition is detected the output of the flip-flop will generate a pulse that can be used to trigger a scope.

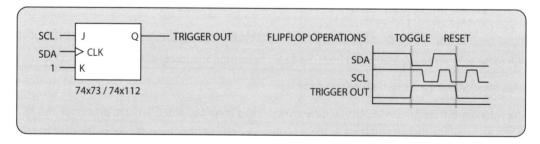

At first glance you may think that the SCL and SDA are connected wrong in the above schematic. Why would you use SDA as clock for the flip-flop? If you recall, the START condition on the I2C bus begins with the SCL and SDA both in a high state. A Falling edge on SDA when SCL is high signifies the actual START operation.

That is exactly what this circuit detects. These flip-flops are sensitive to a falling edge. Only when SCL is high and a falling edge occurs will this flip-flop set its output high. Don't substitute it with just any flip-flop you have laying around. The falling edge sensitivity is the key!

The JK flip-flop toggles state when both J and K are high and a falling edge occurs. This is only the case during a START operation: SCL keeps the J line high and SDA goes from 1 to 0.

The circuit is self clearing. During normal bus operation the SDA line will change state only when SCL is low. So the first negative going transition of SDA after the START will reset the output of the flip-flop again.

This circuit has a little drawback in the sense that the pulse duration is dependent upon the data stream transmitted. If that bothers you the Qn output can be tied to the RESETn input. This will cause the flip-flop to self clear immediately. This pulse will be very short and determined by the propagation delay of the flip flop. You can extend this by inserting a simple r c network or by inserting an additional flip flop that is driven from SCL.

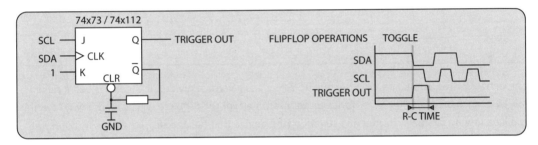

18.2. CHECKING WHO IS CONTROLLING THE BUS

Due to the bidirectional nature of the bus it may sometimes be problematic to find out if a particular device is acknowledging being addressed. While there may be an ACK on the bus it may not necessarily come from the intended device.

A simple resistor can help here. The resistor should be selected to be roughly 1/10 of the value of the pull up resistors on the bus. Simply place the resistor in the SDA line going to the device under test and monitor the normal SDA line (not the line between the resistor and the master)

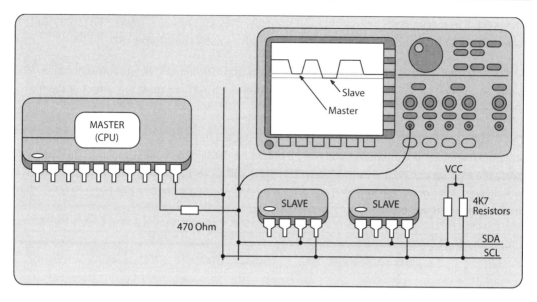

This extra resistor, combined with the pull up will act as a voltage divider. If the bus is being pulled low by a 'normal' device (not the device under examination) the SDA line will go all the way low. If it is pulled low by the device under investigation (the device that has the resistor added in series with its SDA line) the SDA signal will only reach 1/10 of VCC as a logic low level, as the pull up resistor combined with series resistor form a voltage divider.

If you set your scope to 1 volt per division you will have half a division 'offset' on a 5 volts bus, so this will be clearly visible, nor will it disturb normal bus operation since the 1/10 VBus drop is still considered a valid logic 0 on the bus.

During read transactions you can clearly distinguish between address and data fields, and in multi master environments you can clearly see who the master in charge is by applying this trick to each master's SCL or SDA line.

19. I2C INTERFACING SYSTEM FOR THE IBM-PC.

In this section I will set up an environment to experiment with I2C from a PC. The original design used the printer port to perform these operations, therefore for completeness this design is still shown. However, printer ports are a dying breed so alternate solutions are given as well in the form of a USB 'dongle' that can perform I2C operations.

19.1. PARALLEL PRINTER PORT INTERFACE

The schematic is almost the same as the schematic provided by NXP with their older development kits. It is 100% compatible with the NXP driver schematic. However I took the liberty of adding one additional feature: a trigger output to trigger the scope whenever a start is generated. This was done so you can use it both with tools like TV400 from NXP and with my software driver.

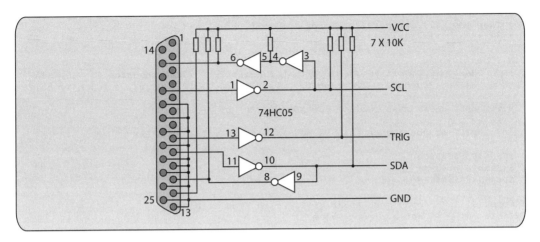

The SCL and SDA signals speak pretty much for themselves. A simple open collector driver stage in the form of a 74HC05 sends the signals out. As said before the TRIG output is an addition I made to the original schematic. When used with the driver described below it is able to trigger an oscilloscope. Each time a START condition occurs the trig output generates a pulse. This will make it easier to monitor bus activity when experimenting with the system. You can write a loop that keeps doing the same over and over again. When you connect SCL and SDA to channels 1 and 2 of an oscilloscope scope and wire the TRIG output to the Trigger input of the oscilloscope you can monitor the transaction without requiring a storage scope.

19.1.1. Software

This module can be used with the programs developed by Philips and available from their FTP site or from the internet. The programs are TV400.ZIP and RAD216.ZIP and they contain a complete bus control terminal. You can monitor bus activity and also actively control lots of standard chips. It comes with libraries for nearly all NXP IC's available. You can also control Non NXP IC's by using the universal bus interface.

19.1.2. QuickBasic / PDS / POWERBasic / Visual basic Driver

This describes a set of routines written in various flavours of Basic that allow you to run an I2C bus over a standard printer port. The program is written in such a way that it can be run on different platforms without (or with minor) modifications.

```
' ***************************
' * I2CDRIVE I2C Bus Driver (c)1995-2010 Vincent Himpe
' * Use as you please. This driver can be used in Quickbasic, PDS, PowerBasic and
' * Visual Basic. Depending on which flavor of basic used, some blocks may need
' * removal.
' ***************************
' * ICee Debugger.
' * If during the running of the program, the computer beeps. This means that an
```

```
' * addressed chip is not responding. Turn on the debugger by issuing the
' * I2CDebugOn command immediately after opening the bus and rerun the program.
' * Then you will get detailed error information
' ****************************

DIM ICdta(10)
COMMON SHARED I2CLoPort,   I2Cmidport,    I2CHiPort,    I2Csavedstatus,   SCL, SDA
COMMON SHARED I2Cdebug,    I2Ctimeout,    I2Cdevadr%,   I2Cresult%,   ICdta(), hold

' powerbasic users must replace 'COMMON SHARED' by 'PUBLIC'

hold =.001  ' do not change this value. Adjust the value in DELAY

' $$$$$$$$$$$$$$$$
' $ Sample program
' $$$$$$$$$$$$$$$$

I2Copen 0, 1     ' open port &h378 (LPT1) with ICee enabled
I2Cinit 10       ' set a timeoutvalue
CLS
I2Cwrite 192, 12   ' write '12' to device with address 192
a = I2Cread(193)   ' read a byte from the same device

I2Cwwsend 200, 1, 7    ' write '7' to register '1' of device at address 200
a = I2Cwwread(201, 1)  ' read register 2 of device at 200
I2Cmultiread 200, 3    ' read first 4 registers (0. . 3) of the slave at address 200
FOR a = 0 TO 3
        PRINT ICdta(a)
NEXT a
END


' ---------------------------
' When using powerbasic you must delete this routine in its entirety. It is
' needed for Qbasic only.

SUB delay (count)
    FOR a = 0 TO (count * 1000)  ' adapt this value according to the CPU speed
    NEXT a
END SUB

' ---------------------------
SUB holdsystem  ' Freezes the system so you can read ICee debugger messages
        CALL messg(" Press key ")
        DO
        a$ = INKEY$
        LOOP UNTIL a$ <> ""
END SUB
' ---------------------------
SUB I2Cclose  ' Closes the I2C driver. This resets the LPT port
        OUT I2CHiPort,I2Csavedstatus
        OUT I2CLoPort,127
END SUB
' ---------------------------
SUB I2Cdebugoff  ' Turns off the I2C debugger
        CALL messg("ICee DISABLED ")
        I2Cdebug = 0
END SUB
' ---------------------------
SUB I2Cdebugon   ' Turns the debugger on
        CALL messg("ICee ENABLED ")
        I2Cdebug = 1
END SUB
' ---------------------------
```

```
SUB I2Cgenstart ' transmits a start
        IF I2Cdebug = 1 THEN
           CALL messg("START")
        END IF
        ' start of transmission
        OUT I2CLoPort, 127
        delay hold
        OUT I2CHiPort, 8
        delay hold
        OUT I2CLoPort, 255
        delay hold
        OUT I2CLoPort, 255
        delay hold
        OUT I2CHiPort, 0
        delay hold
END SUB
' --------------------------
SUB I2Cgenstop ' transmits a stop
        IF I2Cdebug = 1 THEN
        CALL messg("STOP")
        END IF
        OUT I2CLoPort, 255
        delay hold
        OUT I2CLoPort, 255
        delay hold
        OUT I2CHiPort, 8
        delay hold
        OUT I2CHiPort, 8
        delay hold
        OUT I2CLoPort, 127
        delay hold
END SUB
' --------------------------
SUB I2Cgiveack ' Gives an ACK to a slave
        OUT I2CHiPort, 0
        delay hold
        OUT I2CLoPort, 255
        delay hold
        OUT I2CHiPort, 8
        delay hold
        OUT I2CHiPort, 0
        delay hold
        OUT I2CLoPort, 127
        delay hold
END SUB
' --------------------------
SUB I2Cinit (timeout)
        ' aborts transmission and
        ' clears bus
        FOR I2Ca% = 0 TO 5
                OUT I2CLoPort, 255
                delay hold
                OUT I2CHiPort, 8
                delay hold
                OUT I2CLoPort, 127
                delay hold
                OUT I2CHiPort, 0
                delay hold
        NEXT I2Ca%
        I2Ctimeout = timeout
        OUT I2CHiPort, 8
        delay hold
END SUB
```

```
' --------------------------
SUB I2Cmultiread (adr%, count%)
        ' performs a multiread
        adr% = (adr% OR 1)
        I2Cdevadr% = adr%
        CALL I2Cgenstart
        CALL I2Ctransmit(adr%)
        CALL I2Cwaitforack
        I2Cdevadr% = 0
        FOR i% = 0 TO count%
                CALL I2Creceive
                IF i1% < count% THEN
                        CALL I2Cgiveack
                END IF
                ICdta(i1%) = I2Cresult%
        NEXT i1%
        CALL I2Cgenstop
END SUB
' --------------------------
SUB I2Copen (port%, dbug%)
        ' Always call this before
        ' any other routines !
        SELECT CASE port%
                CASE 0
                        I2CLoPort = &H378
                        I2Cmidport = &H379
                        I2CHiPort = &H37A
                CASE 1
                        I2CLoPort = &H278
                        I2Cmidport = &H279
                        I2CHiPort = &H27A
                CASE 2
                        I2CLoPort = &H3BC
                        I2Cmidport = &H3BD
                        I2CHiPort = &H3BE
        END SELECT
        I2Csavedstatus = INP(I2CHiPort)
        IF dbug% = 1 THEN
                I2Cdebug = 1
        ELSE
                I2Cdebug = 0
        END IF
END SUB
' --------------------------
SUB I2Cpob   ' puts bits on the bus
        IF SDA = 1 THEN
                OUT I2CLoPort, 127
                delay hold
        ELSE
                OUT I2CLoPort, 255
                delay hold
        END IF
        IF SCL = 1 THEN
                OUT I2CHiPort, 8
                delay hold
        ELSE
                OUT I2CHiPort, 0
                delay hold
        END IF
        IF I2Cdebug = 1 THEN
                Print "bus acces"
        END IF
END SUB
```

```
' ----------------------------
FUNCTION I2Cread (adr%)    ' reads a byte
        adr% = (adr% OR 1)
        I2Cdevadr% = adr%
        CALL I2Cgenstart
        CALL I2Ctransmit(adr%)
        CALL I2Cwaitforack
        I2Cdevadr% = 0
        CALL I2Creceive
        CALL I2Cgiveack
        CALL I2Cgenstop
        I2Cread = I2Cresult%
END FUNCTION
' ----------------------------
SUB I2Creceive    ' receives a byte
        IF I2Cdebug = 1 THEN CALL messg("Receiving")
        I2Cresult% = 0
        OUT I2CLoPort, 127
        delay hold
        FOR I2Ca% = 7 TO 0 STEP -1
                I2Cb% = 2 ^ I2Ca%
                OUT I2CHiPort, 8
                delay hold
                I2Cin% = (INP(I2Cmidport) AND 128)
                OUT I2CHiPort, 0
                delay hold
                IF I2Cin% = 128 THEN
                        I2Cresult% = I2Cresult% + I2Cb%
                        PRINT "1";
                ELSE
                        PRINT "0";
                END IF
                IF I2Ca% = 4 THEN PRINT " ";
        NEXT I2Ca%
        PRINT
END SUB
' ----------------------------
SUB I2Csettimeout (tme)    ' assigns a timeout value
        I2Ctimeout = tme
END SUB
' ----------------------------
SUB I2Ctransmit (byte%)    ' transmits a byte
        IF I2Cdebug = 1 THEN
                CALL messg("Transmitting")
        END IF
        I2Ca% = 0
        FOR I2Ca% = 7 TO 0 STEP -1
                I2Cb% = 2 ^ I2Ca%
                IF (byte% AND I2Cb%) =_
                        I2Cb% THEN
                        OUT I2CLoPort, 127
                        delay hold
                ELSE
                        OUT I2CLoPort, 255
                        delay hold
                END IF
                OUT I2CHiPort, 8
                delay hold
                OUT I2CHiPort, 0
                delay hold
        NEXT I2Ca%
        OUT I2CLoPort, 127
        delay hold
```

```
END SUB
'   ---------------------------
SUB I2Cwaitforack   ' waits for slave acknowledge
        IF I2Cdebug = 1 THEN
                CALL messg("Wait for ACK")
        END IF
        OUT I2CLoPort, 127
        OUT I2CLoPort, 127
        delay hold
        OUT I2CHiPort, 8
        OUT I2CHiPort, 8
        delay hold
        acknowledge = 0
        acktimer = 0
        WHILE acknowledge = 0
                acktimer = acktimer + 1
                I2C. mon = (INP(I2Cmidport) AND 128)
                IF I2C. mon <> 128 THEN
                        acknowledge = 1
                END IF
                IF acktimer = I2Ctimeout THEN
                        acknowledge = 1
                        IF I2Cdevadr% <> 0 THEN
                                IF I2Cdebug = 1 THEN
                                        CALL messg("No device")
                                        CALL holdsystem
                                ELSE
                                        BEEP
                                END IF
                        ELSE
                                IF I2Cdebug = 1 THEN
                                        CALL messg("No ACK")
                                        CALL holdsystem
                                ELSE
                                        BEEP
                                END IF
                        END IF
                END IF
        WEND
        OUT I2CLoPort, 127
        OUT I2CLoPort, 127
        delay hold
        OUT I2CHiPort, 0
        OUT I2CHiPort, 0
        delay hold
END SUB
'   ---------------------------
SUB I2Cwrite (adr%, dta%)   ' writes a byte to a slave
        adr% = (adr% AND 254)
        I2Cdevadr% = adr%
        CALL I2Cgenstart
        CALL I2Ctransmit(adr%)
        CALL I2Cwaitforack
        I2Cdevadr% = 0
        CALL I2Ctransmit(dta%)
        CALL I2Cwaitforack
        CALL I2Cgenstop
END SUB
'   ---------------------------
FUNCTION I2Cwwread (adr%, sbadr%)
        adr% = (adr% OR 1)
        I2Cdevadr% = adr%
        CALL I2Cgenstart
```

```
        CALL I2Ctransmit(adr%)
        CALL I2Cwaitforack
        I2Cdevadr% = 0
        CALL I2Ctransmit(sbadr%)
        CALL I2Cwaitforack
        CALL I2Cgenstart
        adr% = adr% + 1
        CALL I2Ctransmit(adr%)
        CALL I2Creceive
        CALL I2Cgenstop
        I2Cwwread = I2Cresult%
END FUNCTION
' ---------------------------
 SUB I2Cwwsend (adr%, sbadr%, dta%) ' writes with sub address
        adr% = (adr% AND 254)
        I2Cdevadr% = adr%
        CALL I2Cgenstart
        CALL I2Ctransmit(adr%)
        CALL I2Cwaitforack
        I2Cdevadr% = 0
        CALL I2Ctransmit(sbadr%)
        CALL I2Cwaitforack
        CALL I2Ctransmit(dta%)
        CALL I2Cwaitforack
        CALL I2Cgenstop
END SUB
' ---------------------------
SUB messg (dta$)
        ' shows the ICee messages
        PRINT dta$
END SUB
' ---------------------------
```

19.2. ACCESS BUS

One offspring of the I2C BUS is the ACCESS Bus. This bus was co-developed by Philips (now NXP), Signetics, Digital and Intel. The goal was to create a bus that could help us in getting rid of all cabling involved with computer peripherals. You would have 1 bus connector where you could hook up your keyboard, mouse, digitizer, scanner, printer, monitor etc.

Then by using software you could issue commands to your monitor or printer, things like selecting fonts or programming brightness and picture size could be done from within software on the host platform. Advanced features such as hot plugging are also included in the standard. There has been a chip developed that could replace the standard 8042 (which is the keyboard controller in an IBM PC) by an access bus controller. Digital was using the bus in its Alpha based machines. Sony and some other manufacturers have monitors that support the remote programming features. Microsoft announced support for the bus in Windows 95.

Basically the hardware layer of the access bus is an I2C bus, all that has been done is to implement a software protocol to provide additional functionality to the system.

This bus has never gained any support due to its lack of performance. The idea was good but the speed was not available. Out of this idea the USB was developed. The basic mechanisms in I2C have been modified and taken to a new level. While the physical layer is different, the sync mechanisms, the handshaking, addressing, and multi master mode are an improvement of the I2C bus.

The access buss has never really taken off as it was only suitable for slow user devices such as keyboards and mice. The experience gained from the exercise partially lead to the development of the USB system.

19.3. I2C IN YOUR COMPUTER

You might not realize it but every motherboard using a Pentium II or later processors has an I2C bus onboard! This bus is used for so called system management. This ranges from voltage detectors on the motherboard, temperature monitors in the CPU to the configuration parameters for your memory. Today's DIMM memory modules have a small I2C EEPROM onboard that holds the timing information for the memory. This information is read during boot.

In portable computers the bus is even more used, the little Trackball, Touch Stick or Glide point all use I2C to transmit mouse coordinates! Smart hard disks boast I2C busses. The Battery in a notebook can have an I2C bus that is used to read voltage, temperature and battery information.

This bus is called the System Management Bus and was introduced with Intel's first I/O controller hub chips (ICH 82801). The signals are labelled SMLINK1 and SMLINK0 where SMLINK0 is the clock and SMLINK1 is the data. An interrupt line is present under the name SMBALERT. Both SMBDATA and SMBCLK need pull up resistors just like I2C.

There is no cut and dried method of access to this channel though. Every chipset implements this differently and the access channel doesn't even fall on a fixed address. Low level interaction with the chipset is required to obtain the address of the SMBUs controller and exact layout of the memory space and bus mode is required.

There are some tools for both Windows and Linux that allow the reading of a couple of standard parts like temperature sensors and fan speed. Controlling other devices is not an easy task and the chances are pretty large that you will crash the computer while experimenting with this bus.

Access to SMbus devices is handled through the BIOS and the ACPI mechanism. The BIOS will enumerate all devices and provide a framework to interface to from code. The problem is that you need to provide information to the bios to have it detect your added hardware. Intel and Microsoft have a special compiler that lets you create these entries. And then you need to get them into the bios... not an easy task.

There are ways to bypass the ACPI mechanism and access the SMbus controller directly but here again there is trouble. Each motherboard chipset implements the controller a different way and you need to find out exactly what south bridge chip you are dealing with and then apply the correct operations.

In short it boils down to a task that is difficult to complete without proper documentation of specific hardware. The code is purposed and not easily ported to other hardware and or devices.

For more information on ACPI:
http://www.microsoft.com/whdc/system/pnppwr/powermgmt/ACPIDriver_Vista.mspx

20. COMMONLY USED I2C DEVICES

In this section I will highlight a number of commonly used I2C compatible devices, applications and some practical usages. The examples will be split by classes of devices.

20.1. I/O EXPANDERS

Probably the most commonly used I2C device is an input output device for digital information. The original intent of the bus was to save pins on the processor and avoid complex board routing by offloading the input / output operations to external devices that could be positioned where they are needed.

Since the bus was originally intended for usage in audio and video applications, this is exactly where we find one of the oldest I2C compatible devices: the SAA1300.

20.1.1. SAA1300 5-bit I/O Expander

Even though this device is obsolete it is worth looking into as it has all of the base elements that we will find in a lot of I2C devices.

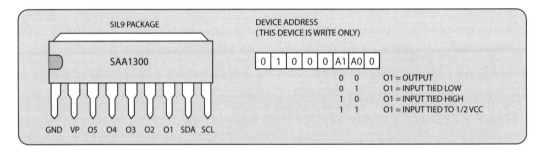

Besides the power and ground we find the SDA and SCL signal lines that are common to any i2c bus device. The device has 5 I/O pins that can be written to using a single byte. The address selection mechanism is a bit odd. Output 1 can act as an address selection input depending on the voltage applied to it at power on. At power up this pin is an address input. If the device is addressed as 0100_0000 then the pin becomes an output and under the control of the data byte.

If the device is addressed at 0100_0010, 0100_0100 or 0100_0110 this pin will remain an input. Essentially you can have a single 5-bit output on the bus by writing to 0100_0000, or you can have three 4-bit drivers by tying the O1 to GND, VCC or ½ VCC and addressing the devices at 0100_0010, 0100_0100 or 0100_0110. Care must be taken in this case not to try to access anything residing at 0100_0000 as this will cause all the SAA1300 chips to fall back into 5-bit mode with potential catastrophic results for those two that have pin 1 tied to GND or VCC.

20.1.2. PCF8574 / PCF8574A 8-bit I/O expander

The PCF8574 is a general purpose digital input/output expander. By setting a combination on the three address pins you can have up to 8 of these devices on a single I2C bus.

The PCF8574 also features an interrupt output. Although this is not I2C related it is a useful feature But since the I2C bus does not really have provisions for interrupt processing you can tie the interrupt output of the PCF8574 to an interrupt input of the processor that acts as bus MASTER.

The outputs are for all practical purposes considered to be open drain. This means they only switch to ground. A weak internal current source can make an output a logical high. However, this current is limited to roughly 100 microamperes, so do not count on it to be able to drive your circuitry.

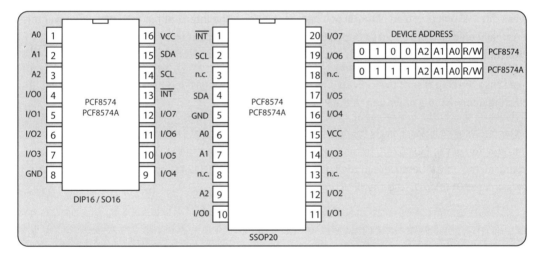

If the PCF8574 is followed by a pure CMOS circuit, or a TTL circuit then it will work correctly. Any other circuitry will require you to add an external pull up resistor or to buffer the signals.

Each pin of the PCF8574 can act as an input or as an output. There is no configuration register to determine the pin direction. Simply setting a pin to logic 1 configures it as input. Since the top side consists of a 100 microampere current source it can easily, and without risk, be over driven from an external source. This mode is called a quasi-input. A pin set as logic high can become an input when overdriven.

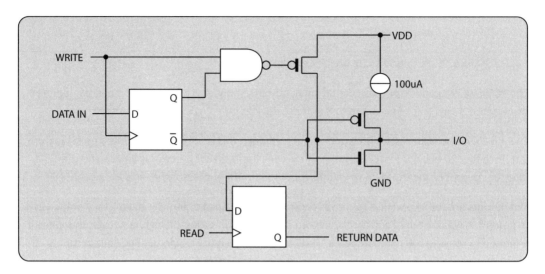

The interrupt output itself is also an open drain construction, however, this one does not have a weak internal pull up, so an external pull up is mandatory. The interrupt pin activates whenever it is externally toggled from 1 to 0 or 0 to 1, but if you set an output pin from 1 to 0 or 0 to 1 the interrupt will not occur.

In a sense the chip is sensitive only to pins that were setup as quasi-inputs. The interrupt is self clearing upon reading or writing to the PCF8574 register.

In larger systems it was quickly found that 8 possible PCF8574 were not enough, hence the introduction of the PCF8574A. In function, pin out and behaviour the PCF8574 and PCF8574A are identical, the only difference is in the I2C base address. The PCF8574 uses 0100xxxD where the A version uses 0111xxxD. (D is the direction bit)

20.1.3. PCF8575 16-bit I/O Expander

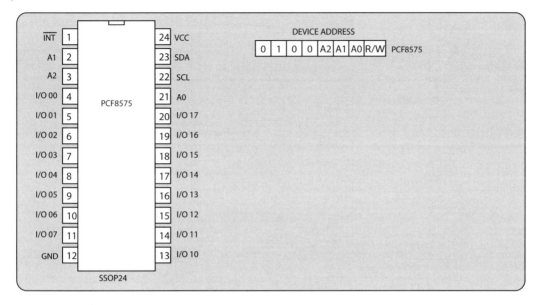

Sometimes more than 8-bits are required and using two devices together is not an option. Take the example where the outputs must remain synchronized with each other. The PCF8575 offers a solution as this device is essentially a 16-bit wide PCF8574. The I/O structure, quasi-input mode and interrupt behaviour are identical. Transfers to and from the PCF8575 should always contain 2 bytes or a multiple thereof. Data will only appear on the outputs when both bytes have been received. This guarantees the outputs remain synchronized in time.

The device address of the PCF8575 is identical to that of a PCF8574: 0100xxxd.

20.1.4. MCP23017 MCP23018

This device made by Arizona Microchip features a 16-bit I/O expander. Similar to the PCF8575 in nature, it has a couple of distinct additions. Amongst those, it features two interrupt output pins that are independently configurable. There where the PCF8574 and PCF8575 have simple I/O registers the MCP23017 has a range of control registers that allow for more complex I/O operations to be performed. The MCP23017 shares the same base address of 0100xxxD with the PCF857x devices, also the internal control registers allow for a fully detailed configuration.

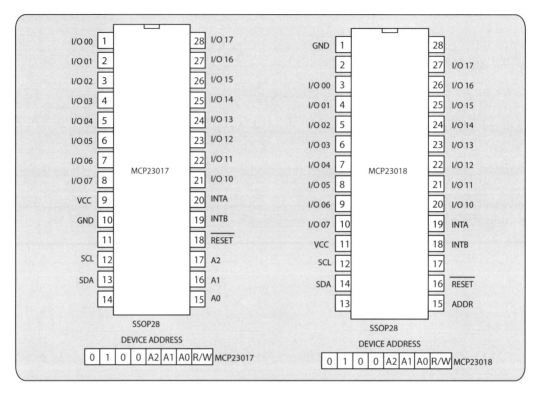

The MCP23018 is nearly identical to the MCP23017 but has open collector outputs and a different pin out. The MCP23018 has a unique way to select the address. The address pin is actually an A/D converter that digitizes the voltage at this pin to allow the selection of one of 8 possible sub addresses. These devices are also available in 8-bit wide variants.

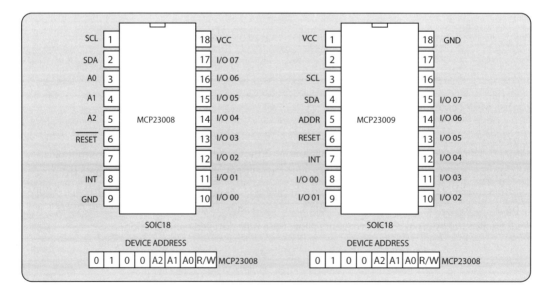

The MCP23009 employs the same analog address selection mechanism as its MCP23018 counterpart.

20.1.5. Non-volatile Expander

There are I/O expanders available that have the capability to retain their settings in non-volatile memory and to restore the setting upon power up.

The Maxim (formerly Dallas Semiconductor) DS4520 is such a device. Besides the 8-bit I/O port it also has 64 bytes of EEprom onboard. The I/O memory is also special since it is essentially an EEprom shadowed by a RAM. This means that the user can randomly read and write in the RAM page and store the data into the EEPROM by issuing a command. This greatly reduces the normal wear and tear on EEproms that are frequently written to. During normal operation the I/O state can be changed as often as required without incurring an EEprom write.

The PCA8550 is a rather unique device that was originally developed for a control function on Pentium ™ processor motherboards. The device has a 5-bit memory. 1-bit controls OUT0 while the other 4-bits can be made available on the MUX outputs. An external pin (MUX SEL) selects whether the MUX IN signals are connected to the MUX OUT signals, or the 4-bit memory is connected to the MUX OUT signals. This allows switching between a live signal and a stored signal on those 4 output pins.

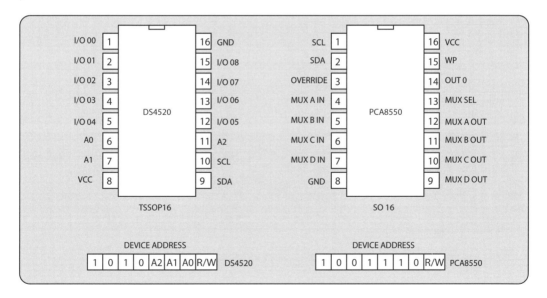

20.1.6. Other Expanders

There are many I/O expanders' available on the market and later on in this book I will give more examples of such expanders as used in a variety of applications. Many manufacturers make different

I/O expanders for specific applications; therefore a simple search on I/O I2C should yield a bunch of results. The largest one being a 60 port expander made by Cypress: the CY8C9560A.

20.1.7. Applications

Anywhere a digital signal is used these expanders can be utilised. From simple things like driving LED's or relays, sensing quasi-static signals, to interrupt driven applications such as keyboard interfacing, can all be done using these universal components. The advantage comes from not having to route all these signals throughout your system. You place the expander chip exactly where it is needed and only the SDA and SCL signals need to propagate through your system. This saves a lot of routing real estate on the board, it also frees up a lot of valuable I/O pins on your main processor. Some I/O expanders carry enough intelligence to offload part of the work of the main CPU.

When picking an expander for a particular task, look at exactly what you want to do and then match that to one of the available devices. There is plenty of choice and you don't need to be stuck with a PCF8574.

20.2. LED AND LCD DISPLAY CONTROLLERS

There are numerous LCD display controllers available for the I2C Bus. These devices offer an extensive capability to drive multiple display and large segment counts. Driving an LCD is not an easy task since it requires applying an AC Signal to the segment electrode and the backplane electrode. You cannot use a DC signal since this causes electro migration of the electrode material and would destroy the LCD over time. The positive electrode would be gradually 'eaten' away. By using an AC signal this problem is alleviated. The drawback is a complex drive waveform.

The waveform is not easy to understand. Let's first take a look at a so called static LCD display. (Left image)

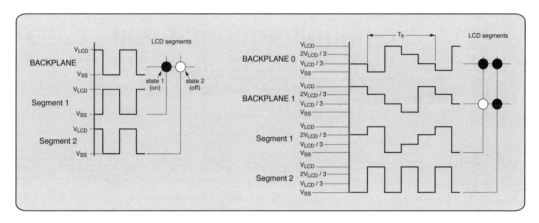

Each pixel on an LCD consists of 2 electrodes facing each other. One electrode sits on the bottom glass sheet, the other on the top glass sheet. In between the glass plates sit a layer of liquid crystals. To reduce the amount of electrical connections they connect a group of the electrodes of the bottom glass plate together and call it a 'backplane".

So a group of pixels share a common electrical contact leading to one of their electrodes. The front electrodes for each pixel are brought out individually; these are the so called driving electrodes. We apply a 50% duty cycle square wave to this common point. To turn on a segment we need to have a voltage differential across the plates of a pixel (State 1). If we apply a signal to the backplane, in anti phase to the driving electrode we will have a voltage differential across the pixel, making it visible. If we apply to the backplane a signal in-phase to a pixel there will be no delta voltage and the pixel will remain transparent (state 2).

To drive a lot of pixels we can make a matrix of the bottom and top electrodes (right image). The driving waveform becomes quite complex. For a given point in time the voltage differential is important. If none or only a very small differential exists the pixel will be off. If a large differential exists the pixel will be black (on).

The LCD controller can handle up to 4 backplanes with varying pixel counts. Needless to say that such a waveform looks very complex. The LCD controller creates all the drive voltage levels and the necessary timing to scan the array of pixel elements.

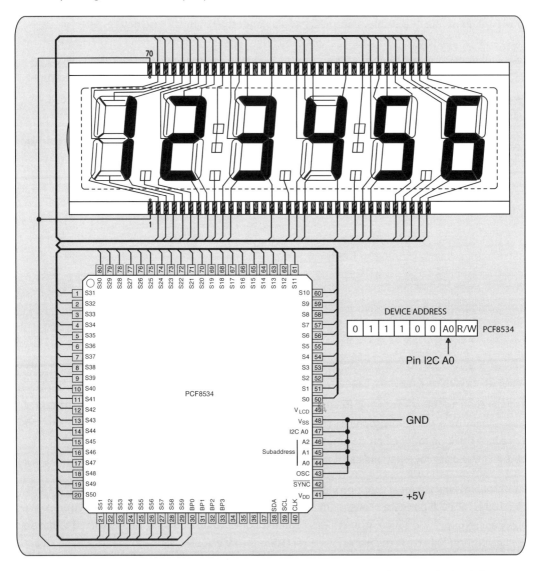

Above is an application that uses a PCF8534 in static operation mode. In this case the backplane connects all segments and is driven by a simple 50% square wave. I connected a standard 6-digit 7-segment display. It is perfectly possible to connect 3 more of these displays., all I need to do is to connect the driving electrodes of each display together (all digit1 segment A electrodes go together, all digit1 segment B electrodes go together etc) and then feed the backplane of each display with a different backplane signal (BP0 to BP3).

The display controller has enough memory to remember the state of each pixel and I can connect up to 16 of these display controllers together in a variety of backplane drive modes. The controllers can run in unison with each other by connecting the SYNC and CLK pins together.

The addressing is a bit unique as there are only 2 I2C addresses, but each address can have 8 devices giving a total of 16. The first byte transmitted after the address byte holds 3 more bits that select 1 of 8 devices. The I2C A0 bit selects which 'bank' of 8 you will talk to, the sub address lines A0 to A2 in combination with the transmitted byte pick which of the devices in a bank you actually talk to.

These controllers do not have character generators and it is up to the user to create the correct bitmaps to render an image.

20.2.1. LCD Display Modules

Some manufacturers have created I2C compatible modules that behave like the well known HD44780. These modules employ an ASIC that has a similar behaviour and instruction set but replaces the parallel bus with an I2C interface. An onboard voltage generator creates the necessary drive voltage but can also be controlled over the I2C Bus. The contrast is software programmable.

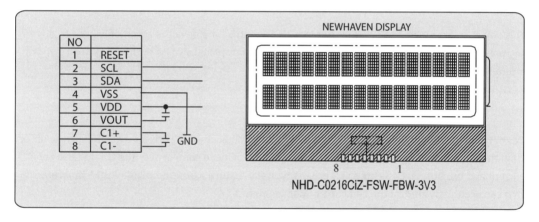

These modules often come in so called COG or 'Chip-On-Glass'. The controller ASIC is a bare piece of silicon that is bonded directly to the glass of the display. Even the contact pins are glued directly to the glass plate of the display. The advantage is a very thin display. Even backlight is an option. The image above shows such a display that has a COG construction and also has a backlight.

20.2.2. LED DISPLAY Drivers

Besides the LCD driver chips and modules there are also plenty of LED driver chips out there that sport an I2C interface. Probably the best known is the SAA1064.

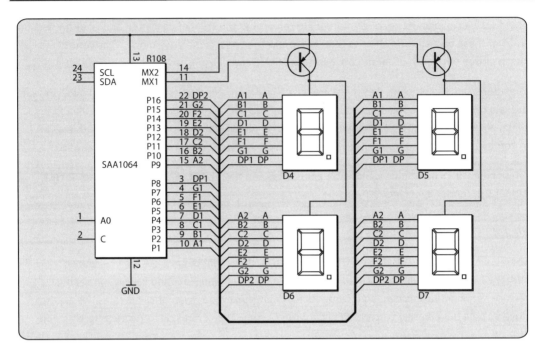

This device can manage 32 LED's that run in a 2 x 16 matrix. The device is commonly used to drive 7 segment displays but can be used for almost any kind of display. Later on in this book I will give you a complete example using this device.

20.2.3. LED Drivers

Besides pure display drivers there are also a number of dedicated LED drivers available. These differ from the display drivers in the sense that they offer much more control over the LED in terms of current and brightness. Dedicated LED drivers typically feature a programmable PWM generator that can accurately control the Led current per segment.

There are two principal categories of devices. There is the full control LED controller such as Texas instruments TLC59116 and NXP9635. These devices allow full control of intensity and blink control per LED.

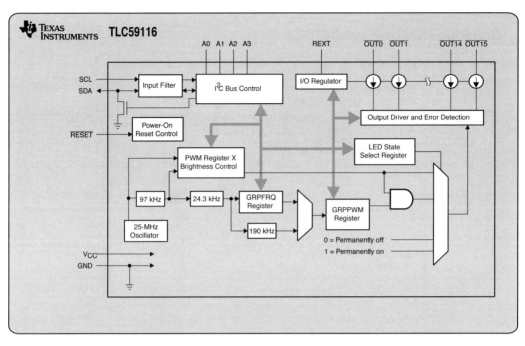

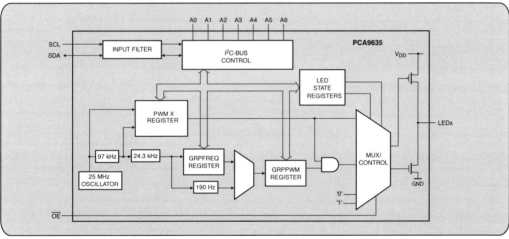

And then there are the so called LED blinkers such as the PCA9532. These devices only allow for 2 settings and the user can select which settings go to which LED. A LED can be ON, OFF or driven from either Rate1 or Rate2. There is no simultaneous individual control for each LED, but these controllers are ideal for indicators since they can autonomously make LED's blink.

LabWorX   Commonly used I2C devices

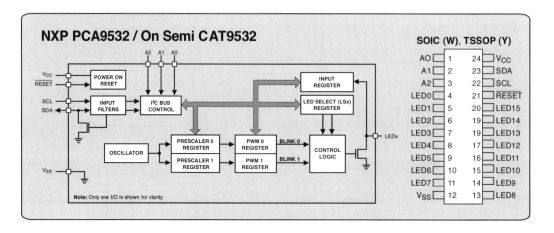

20.3. A/D AND D/A CONVERTERS

There are plenty of devices available to interface with the digital world, but what do you do when you are dealing with the analog world? We need an interface layer that can quantify an analog signal and return us a numerical representation. After processing we may need to convert our calculated quantity back into an analog form, as either a voltage or a current.

Cue the analog to digital and digital to analog converters.

20.3.1. Principles of Digital to Analog Converters

Before we take a look at the digitizing (analog to digital) portion, I will first cover the inverse process: digital to analog conversion. The reason is that some A/D converters are actually DA converters in disguise.

20.3.1.1. THERMOMETER DAC

The simplest implementation of a Digital to Analog Converter (DAC) is the so called thermometer scale DAC. A simple decoder closes one of a possible range of contacts. A voltage divider creates a range of subdivisions of a reference voltage. The divider chain is connected between a Vref and ground, or both ends of the divider chain can be brought out.

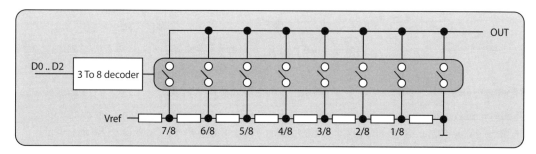

The problem with this kind of DAC's is that the error that is introduced when a resistor deviates from the required value. All resistors must be identical for this DAC to work correctly. Any deviation will propagate throughout the whole chain and cause errors for all outputs. This makes this DAC difficult to construct when any kind of precision is required. For simple 4-bit DAC's this is still possible, though above that the linearity very quickly degrades.

An 8-bit thermometer DAC would require a chain of 257 resistors and 256 switches. This combined with the necessary decoder logic requires substantial surface area on the integrated circuit and makes these DACs impractical.

Another problem is that the output impedance of this DAC changes depending on the setting. These DACs should always be combined with a buffer amplifier (voltage follower).

20.3.1.2. SWITCHED RESISTOR NETWORK DC (R-2R)

Another type of DAC is the R-2R network DAC.

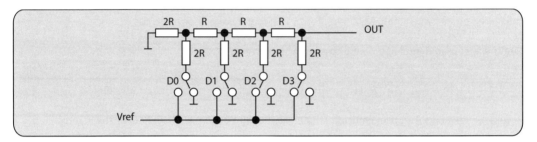

This construction greatly reduces the amount of resistors in the chain. Only two resistors per bit are required. A couple of transistors allow the switchover from ground to Vref.

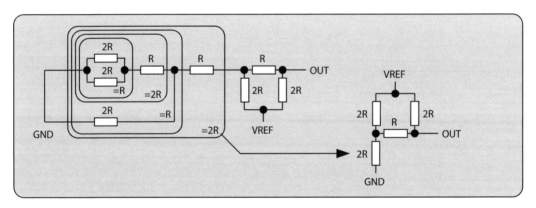

The principle is simple and illustrated above. Depending upon the position of the switches you create a divider between Vref and GND. It is important for the switch resistance to be as low as possible since this has a direct impact on the output voltage. Just as with the thermometer DAC, these converters need an output buffer amplifier.

This kind of converter is cheap to manufacture and can be made reliable with good linearity and precision.

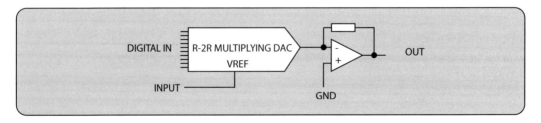

The advantages of Thermometer and R-2R DAC's is that they are basically switched resistor chains. This means you can use any kind of signal to drive Vref. On some DAC's the Vref pin is actually brought out. You can even send AC signals superimposed on a DC level through there. There are applications where these DACs are used as the feedback resistor in an opamp circuit to form a programmable gain amplifier or filter.

These DACs are sometimes referred to as multiplying DACs, specifically for the capability to create a fractional value of an incoming signal.

20.3.1.3. VOLTAGE OR CURRENT SUMMING DAC

A different principle is the summing DAC. The previous DAC systems had drawbacks in terms of linearity and precision due to the injection of resistance in either the switching element or simply in the difficulty to create matched resistances over a large range.

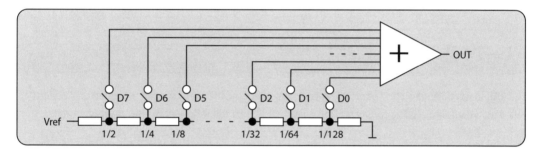

Summing DACs solve this problem. The resistor chain is simple to construct and only has a few elements. A voltage or current summing amplifier adds the selected partial currents and creates the desired output level. Since only few resistors are used these converters are simple to construct. The series resistance in the switching element has no impact on the overall performance. The resistor pattern can easily be created so that it allows for factory trimming. The absolute value of the resistors is unimportant. Only the relative values are important.

20.3.1.4. PWM OR BRM DAC

Pulse width Modulation or bit Rate modulating DAC's are a different kind of digital to analog converter. As opposed to a system where analog signals are summed or switched under digital control, these circuits create a time related output signal. After integration of this signal over time an analog output signal is reconstructed. The precision is purely dependent on the accuracy of the integrator. The base frequency is unimportant since the output signal is determined by the ratio between high and low. Even if the frequency changes this ratio remains set. The whole converter remains digital up till the point of integration.

There are various implementations possible for the integrator and this can go from a single R C network to a switched current source implementation.

Current consumption of such devices can be very low.

20.3.2. Principles of ANALOG to Digital Converters

Now that we have seen how a basic digital to analog circuit is constructed we can take a look at the various digitizing processes for an analog signal.

20.3.2.1. SUCCESSIVE APPROXIMATION

The Successive Approximation Register or SAR is not a converter in itself but part of an analog to digital converter system. However, in practice the SAR moniker is used to indicate complete systems that employ a SAR system to perform the conversion.

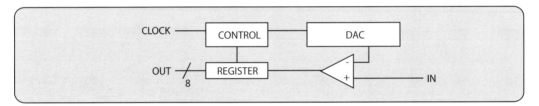

The SAR in itself is not a converter; it is merely a control mechanism to build a converter. The SAR relies on a digital to analog converter and a comparator to tell it how accurate its guesswork is.

The SAR register is a state machine that tries each bit once and verifies if that bit represents a higher or a lower voltage than the one that needs digitizing. This mechanism is also called a binary search. The SAR will start at zero and it will change its most significant bit (MSB) to 1. The output of the SAR goes to a D/A converter that creates an analog value representing the contents of the SAR. An analog comparator looks at the incoming signal and the signal generated by the DAC. The output of the comparator is fed back into the SAR. If the comparator yields logic 1 the bit is left set to 1, otherwise it is cleared. The SAR then proceeds to do the same action on the next significant bit (MSB 1). This continues until all bits have been tested and compared. The final contents of the SAR now hold the digitized value of the input signal.

This kind of conversion is done under the control of a clock and the input signal must not change during the conversion. The precision of the SAR is determined by the precision of the DAC system used. The SAR system can be extended to any number of bits required but conversion speed will drop proportionally with the number of bits to be digitized.

Since in this converter system a D/A is present this can be used as a bonus. The PCF8591 / PCA9691 are SAR based converters. When not converting the D/A is used to create analog outputs.

20.3.2.2. FLASH CONVERTER

To overcome the slowness of the SAR a different mechanism can be employed. Just like with the thermometer DAC a resistor chain of identical resistors creates a set of reference voltages. Each reference voltage is compared to the incoming voltage using a comparator. The highest comparator in the chain that has its output set, thereby indicating the level of the incoming signal.

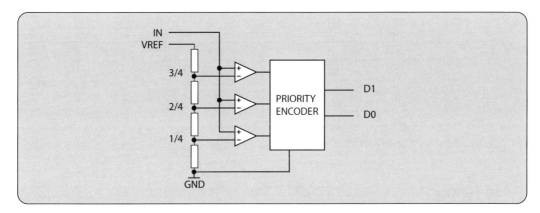

A priority encoder picks up all these comparator outputs and produces the binary number; the advantage is that the conversion is instantaneous. Alas, the drawback lies in the precision of the reference divider chain, the number of comparators required, and the area required to build such a large priority encoder. For an 8-bit flash converter we need 2^8 or 256 comparators.

A variation is a so called pipelined flash converter. In this system a number of flash converters create a seek range. A 2 or 3-bit converter may take an instantaneous reading of the range where a signal falls. A second flash converter in the chain is then switched to digitize an additional two or three bits within that range. This can be followed by additional converters to increase the range.

Even though the conversion is now driven by a clock to hand off from one sub converter to another the speed can still be very high. Every clock tick takes in a new sample. Even though it may take multiple clock ticks for the sample to be fully digitized, the throughput happens at full clock speed. The principle is that of a production chain. Even though it takes 8 hours to build a car a chain will yield hundreds of cars a day.

20.3.2.3. SINGLE SLOPE AND DUAL SLOPE INTEGRATING CONVERTERS

The integrating converter offers excellent precision and resolution. Most notably the dual slope converters are used extensively in test equipment, often yielding results in the order of 18 to 24-bits and more. The ICL7106 and 7109 are well known implementations of dual slope converters.

The conversion is purely time related and solved in the digital domain. The only analog element is a capacitor and a reference current.

Whenever a conversion is initiated the capacitor is fully discharged. The current source is then turned on and the time is measured to reach a voltage across the capacitor that is equal to the input voltage. In fact this thing acts as an electron counter.

The only source of error in this system is the capacitor and the reference current. To solve this problem the dual slope system was created. This mechanism eliminates even these errors.

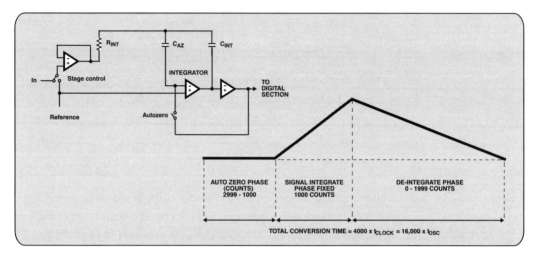

In a dual slope system the capacitor is first pre charged using a fixed time from the incoming voltage (the voltage to be digitized). The absolute value of the capacitor is irrelevant. After this fixed time the capacitor is now discharged using a reference current until we hit 0. The time required for discharge is a direct measure of the input voltage.

An auto zeroing system can be constructed to 'remember' any residual charge or offset in the system. This is simply stored as a voltage across the capacitor before the sequence starts. Any error present is then automatically accounted for.

Even if this system drifts under the influence of temperature, it will self-balance since the conversion rate is much higher than the thermal reluctance.

The drawback is that the conversion time is not constant. The conversion ends whenever the end value is reached. The required time fluctuates depending on the incoming voltage. These converters are typically slow (a few readings / second) but very accurate.

20.3.2.4.   SIGMA DELTA CONVERTERS

A sigma delta converter is the opposite of a PWM DAC. This is a very complex circuit to build and suffers from a range of difficult to overcome problems. The principle is simple though.

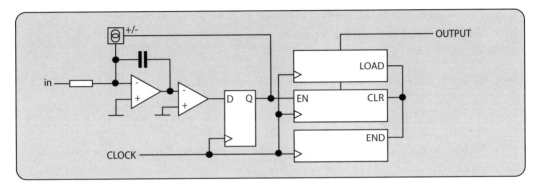

Take an incoming analog signal and use it to charge a capacitor (integrate it). Compare this charge with a reference. Depending on if it's higher or lower we will turn on a current source to charge or discharge the capacitor. In essence we are creating a PWM signal stream. The duty cycle is a measure for how far we are 'off' from having a balanced circuit.

A digital block now looks at this PWM signal and performs a duty cycle calculation in a given time frame. The output of this duty cycle calculation is the digitized value.

The problems are time related. If an incoming signal has a frequency that is higher than the runtime of the calculation, this will completely mess up the conversion. Filtering of the incoming signal is required and a sample and hold system is mandatory.

However, the analog portion is very small in this kind of converter and well under control. These converters can be made cheaply because they require little chip real estate. These are the favourite converters to embed in many microcontrollers just for that reason.

20.3.3. Things to Be Aware Of:

When dealing with converters you will frequently encounter terms like accuracy and resolution. These may be confusing and are often misunderstood.

20.3.3.1.   ACCURACY AND RESOLUTION

Resolution is an indication of the smallest unit that can be measured. For example, an analog clock with only an hour indicator has a resolution of 1 hour (assuming that we don't read in between marks, so assume the needle jumps only from hour to hour).

We can measure in increments of hour, therefore the resolution of this clock is 1 hour.

Accuracy is how much the system deviates from the true value. If it is 12:30 the clock is inaccurate because its resolution does not allow subdividing between hour marks. Even a broken clock (stuck needle) is accurate twice a day! As time passes on and the true time begins deviating from what the clock shows the accuracy drops very quickly.

A system can never be more accurate than its resolution. Even though a converter may have a resolution of 1 microvolt, there is maybe so much noise and drift in the system that the accuracy is +/- 100uV.

The better the accuracy of a system the smaller the margin of error is.

20.3.3.2. DNL AND INL

Differential and Integral nonlinearity are a measure of how good an A/D converter performs.

20.3.3.2.1. DNL

DNL is a measure for how good a step approaches the value of 1 LSB. This may sound like a complex item but actually it isn't.

Let's assume a converter that has 1 volt per step. Ideally a voltage of 0.999 would still read as 0 while a voltage of 1.001 would read as 1. Similarly a voltage of 1.999 would still read as 1 while a voltage of 2.001 would read as 2.

In reality there will be an offset and drift. It may be that we actually need to go above 1.1 volt before it will register as 1, while a voltage of 1.950 may already register as 2.

Ideally the code should only change every one volt. To change code from 1 to 2 it took only 850 millivolts: we were at 1.1 to obtain the code for 1 and we needed only 1.95 to obtain the code for 2. The increment was 1.95 – 1.1 or 850 millivolts.

This differential between the actual increment needed and the theoretical increment (a step equivalent to the value of 1 LSB) is the DNL. Keep in mind that DNL is not a fixed number! It is a diagram that shows the trend for each step. Under no circumstances should the DNL go above ½ LSB or below – ½ LSB.

DNL errors are typical noticeable when large steps are taken (like rolling from 01111111 to 1000000). The reason is that a new voltage or current source is thrown in the mix and the accuracy of this source may deviate from the combined accuracy of all other sources that were active.

Ideally DNL plots should be given once for an upward sweep and once for a downward sweep. The reason is that the comparators in the A/D converter may have hysteresis and this will have a direct impact on the DNL. If this hysteresis is asymmetrical then the values for DNL will be different depending on if the step was taken going up or going down.

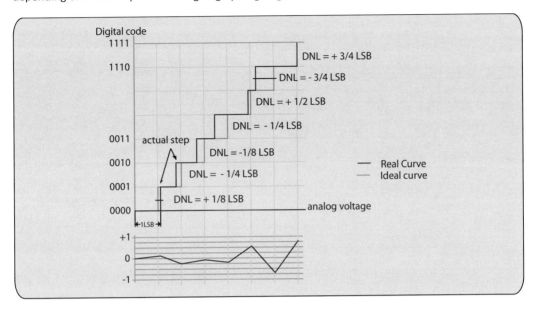

The above diagram shows the tripping points for a real converter (black curve) as opposed to the ideal converter (grey curve). The deviation is either positive (trips too late) or negative (trips too early). The trend is plotted on the bottom curve. The flatter this curve, the better the performance of the converter.

Remember that DNL is differential; therefore you need to take the trip point of the current code and compare it to the trip point of the previous point. The voltage increment required should be exactly the value of 1 LSB. Any deviation is marked as DNL.

DNL only gives only an indication of the monotonicity of the steps. To find out how far the actual trip point lays from the ideal point you need to look at the integral non linearity.

20.3.3.2.2. INL

Just like DNL the INL is not a fixed value but a curve. Ideally it should be as flat as possible

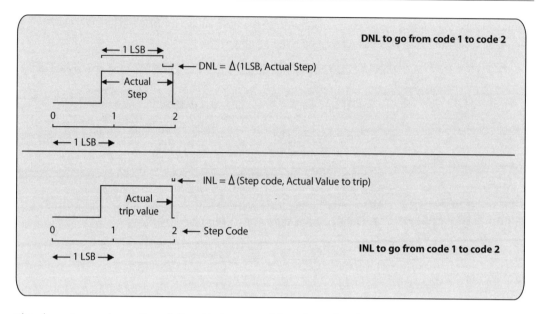

The above image shows the relationship between INL and DNL for the same step.

When evaluating an A/D converter you need to look at both DNL and INL. It is perfectly possible for an ADC to have good DNL but really bad INL. Assume the converter always trips 100mV too early. The DNL will still be 1 LSB so it will read 0, but the INL will be off by 10% for every step!

Manufacturers publish these curves for their high performance converters or they will simply state what the max DNL and INL is across the scale. Some converter types have excellent linearity.

20.3.4. Practical Converters with an I2C interface

There are literally hundreds of converters available with an I2C interface. The advantage being that I2C only requires 2 signals, which allows the placing of the converter directly where it is needed. You no longer need to run analog signals over a long distance to get them to a digitizer, nor do you need to make room for a large parallel bus that runs throughout the board.

Converters are available in a wide range of bits, speeds and channels counts.

Some of these converters come in extremely small packages like a 5 or 6 pin sot-23 or sc70. This allows the user to conveniently scatter these around wherever they are needed.

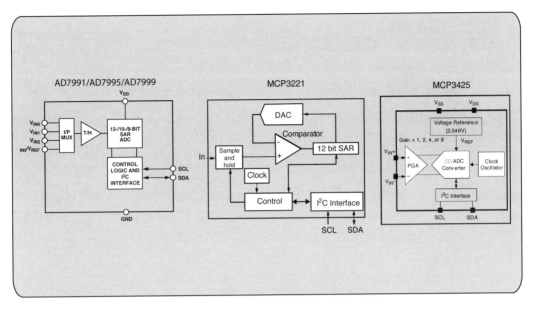

Later in this book I will present various modules to perform D/A and A/D operations and their applications.

20.4. AUDIO AND VIDEO CIRCUITRY

Even though the I2C bus finds its origins in audio and video applications, little remains of those original circuits. In the era of analog signal processing for radio and TV the I2C bus was introduced to allow easy digital control of the analog signal path. There were hundreds of circuits from tuner, mixer, and demodulator over signal multiplexer, teletext and closed captioning circuitry to complex audio decoders, band filters and control amplifiers. These have mainly gone the way of the Dinosaurs since the processing of these signals has changed to digital. With the advent of Plasma and LCD TV the classic circuitry has been discontinued and only a few circuits remained. Most manufacturers of components in this field have long since discontinued their devices and NXP was the 'last man standing'. In March of 2010 the last wafer fabrication making these older devices was shut down and dismantled. In one instant almost all these devices ceased completely to exist.

There are still audio circuits in mass production and new devices are being added periodically.

The prevailing devices are now power amplifiers that offer I2C control.

A lot of those devices target mobile applications such as cell phones and very specific device applications. However, in the car audio and home entertainment class there are some interesting devices to experiment with, like the TDA7561.

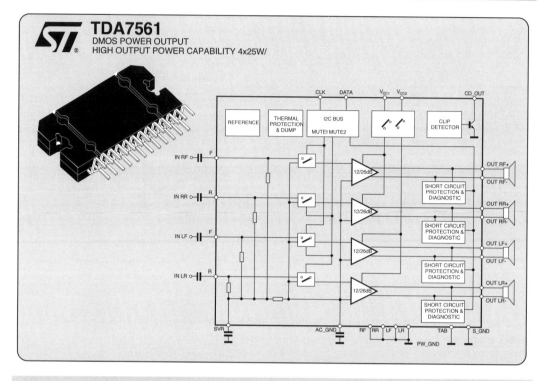

20.5. EEPROM MEMORIES

The second most used device on the I2C bus is probably a type of E2Prom (EEPROM). Since the bus was intended for audio and video applications, such a memory would be useful to store the user presets for audio as well as tuning information. The PCF8581 and PCF8582 were among the first E2Proms available.

The PCF8581 soon became obsolete (it was actually a PCF8582 with some bit defects and the faulty page disabled). Due to a growing interest in these memories multiple manufacturers started making functional clones, some even without having an official I2C license. These devices were sold as '2-wire' or '2-wire compatible' E2Proms. One notable manufacturer was Xicor (now a part of the new Intersil).

Xicor launched the 24cxx series of E2proms. Over the years the capacity grew from 128 x 8-bit (24c01) to 2048*8-bit. The base device 24C02 was a direct clone of the PCF8582. This 256x8 device uses an 8-bit sub address to select a byte in the array. Larger devices simply use multiple I2C addresses to select a 'page'.

A 24C04 device has two I2C addresses: 0101xx0D and 0101xx1D and it behaves as if there were two 24C02 devices on the bus, sitting at a consecutive address.

The 24C08 and 24C16 behave as 4 and 8 24C02 devices.

The drawback is that the hardware address pins were sacrificed. There where you can put up to 8 24C02 devices on a bus, you can only put 4 24C04 devices on a bus.

This scheme could not continue for larger devices and thus a different solution was introduced, starting with the 24C32 the sub address is a 16-bit word, i.e. two 8-bit transmissions. This allows for a 65535 x 8 matrix. The 24C32, 24C64, 24C128 24C256 and 24C512 devices use this addressing scheme. Once this scheme was exhausted, they had to fall back to the original mechanism. The 24C1024 again listens to two base addresses, each sub address containing half a megabit.

The 24Cxx series is available from multiple manufacturers, spanning a wide range of operating voltages and I2C speeds. Some devices even have a hardware write protect pin. Most of the devices are pin compatible, although one or two devices exist with a mirrored or rotated pin out.

Using the larger memory is simply a matter of changing devices. As long as you stay within one family (24C01 to 24C16 is one family. 24C32 to 24C512 is the next family) your software code remains largely the same.

There are a couple of caveats though. The smaller devices generally can only write up to 4 bytes at a time. And these bytes must reside in the same 'page' of memory. The background reason is the operation of an E2Prom cell. An EEprom cell can only be written to logic 0. When the cell is erased it is set to logic 1. The erasing process takes a long time, so to compensate for that process, they use a 'bulk erase', in other words; multiple cells are erased in parallel.

The smaller EEPROMs have a buffer memory for 4 bytes. When you want to 'burn' data into the EEprom, the current data is copied to the buffer, the buffer is changed with up to 4 new bytes, the block erase command is executed and the 4 bytes are reprogrammed.

For the larger E2Proms this page size has been increased to speed up the programming process further. Depending on the type the buffer can go all the way up to 256 bytes.

This has an important consequence. The number of write operations to an E2prom cell is very limited. Every write operation leaves a permanent 'scar' on the barrier material, eventually this material will fail and the cell will not be writable anymore. Because erase and programming happens in 'pages' this means that all the cells within a given page age the same.

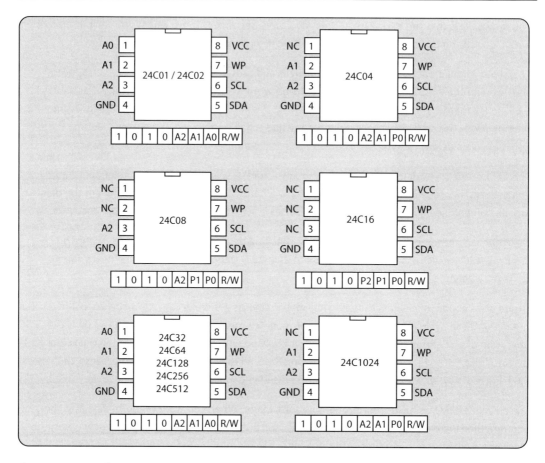

If you are aiming for maximum reliability you should store the information redundantly between pages. By alternating between writing page x and page y you can double the reliability of the data. For even more reliability you can cycle over multiple pages. Since a page contains multiple bytes you can insert a sequence number or you can erase the previous page as well.

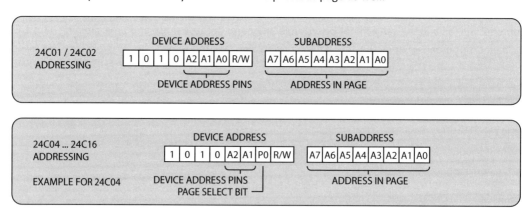

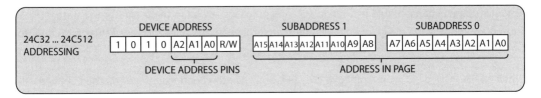

The reading operation then only has to look for the gap and grab the next page in sequence to restore the latest data.

20.5.1. E2Prom Technology

A basic EEPROM cell is a MOSFET transistor that has two stacked gates. The channel between source and drain is covered with a Silicon oxide layer. On top of that is a first gate structure. This gate in turn is covered with another oxide barrier (oxide is non conductive) and the second gate is applied on top of the first.

These two stacked gates form a capacitor.

During programming a current is pulled from Source to drain. The electrons running through the channel are suddenly exposed to a higher electrical field present on the top gate and they jump the barrier (called tunnelling) to land in the floating gate thus charging this. This charged gate changes the threshold voltage of the MOSFET.

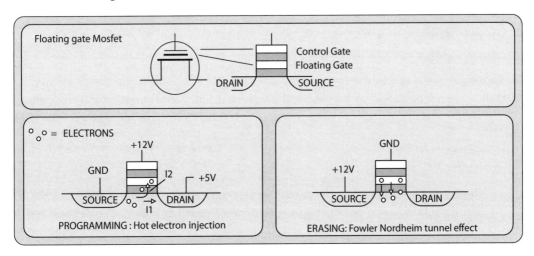

A charged gate changes the normal threshold voltage required to turn the MOSFET on. To erase the cell the polarities are reversed and the electrons jump the barrier again, this time discharging the floating gate.

The problem is that whenever these electrons jump the barrier they can get trapped inside this isolation barrier. This eventually leads to a degradation of the cell and leaves the EEPROM cell stuck in the programmed state. In the beginning, this would happen after only a few hundred cycles. The residual trapped charge would be then large enough to lock the bit in the programmed state. Subsequent improvements in technology and structural geometry have increased this cycle count to well over 100K cycles, but they too will eventually wear out.

When an EEPROM bit is programmed, a charge is stored in the floating gate. When the EEPROM cell is read the transistor is switched on yielding a logic zero. An erased cell has no charge and the transistor is switched off, therefore it returns a logic 1.

20.5.2. FRAM Devices

The FRAM is an outwardly compatible device to an EEPROM. The memory structure is different though, instead of the classical floating gate MOSFET used in EEPROMs, the FRAM uses a Ferro electric material to store information in an electric field.

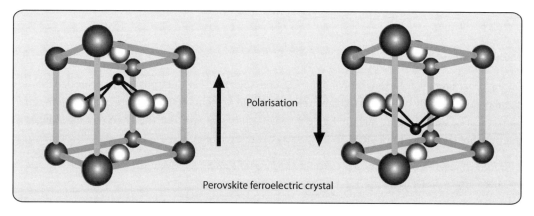

Perovskite ferroelectric crystal

The Perovskite crystal structure traps an ion inside a crystalline structure. By applying an electric field this ion can be pushed up or down inside the lattice and this changes the electrical properties of the molecule. It is this underlying change that is used in Ferro electric memories to store the state of a bit.

The advantages are numerous

- The write and read speed are the same. There is no erase cycle, so there is no problem with wearing out the device. You can perform random writes without being penalized ever in a lifetime.
- The devices don't need to generate the high voltage required for erasing, so this gives lower power consumption.
- There is no wearing out of the memory cells because of charge injection since this device is not charge based.

Do not confuse the term Ferro electric with Ferro magnetic. There is NO magnetism involved in these devices. The storage element is still a capacitor, but one made with a ferroelectric material and not with a floating gate like in the classic EEprom (E2prom).

An FRAM has a much faster access, where a classic EEprom may require a few milliseconds to write a single page the FRAM can be written completely in this time. Due to this speed there is no need to wait for write operations to complete; in fact, the FRAM itself is much faster than the I2C bus.

The parts are usually a simple 'drop-in' for classic EEprom devices and will work without any code modification.

However, if you want to make use of the enhanced speed now possible, you can eliminate any paging mechanism you implemented in code. These devices do not use a page system internally. You can place the address pointer anywhere and just keep sending bytes. The address pointer will simply increment until it hits the end of memory and then roll over to zero. This can give a reasonable speed boost as there is now no need for the code overhead to compensate for the paging limitations and/or wait time for write cycles to complete.

20.6. REAL TIME CLOCKS

Real time clock chips are available in all sorts and flavours with the PCF8573 and PCF8583 being among the first of them. Some devices only offer simple timekeeping while others feature an oscillator output, wakeup signal or alarm function. Some devices even have extra storage space available that is protected by the battery backup.

RTC devices have one thing in common: they operate over a wide voltage range, and draw a very low operating current. When they detect a low voltage these devices will either switch over to battery operation mode (if a dedicated VBAT pin is available like the ISL1208 device) or they will stop responding to I2C operations.

Almost all of these devices are driven from a 32768 Hz crystal although some devices run off a 32000 Hz crystal or can even run from 50 or 60Hz inputs.

Contrary to the E2prom devices there is no real fixed pin outs for the real time clocks. There are hundreds of devices out there, each with their own feature list.

Here is a short grasp of what is available and what sets them out from the others:

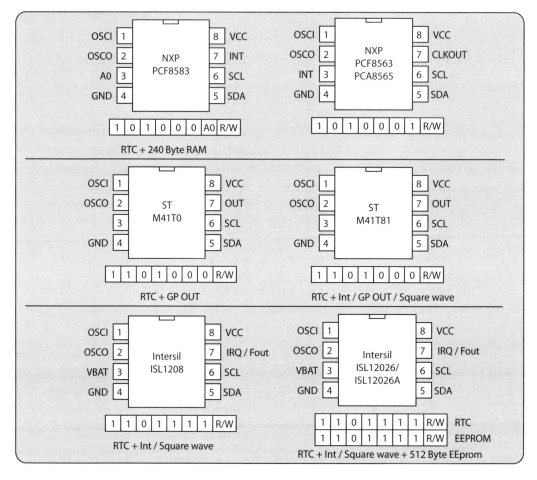

There are many more like the Dallas / Maxims DS1337 that has two interrupt outputs that are fully programmable. Almost any IC manufacturer that has I2C devices has one or more Real time clock circuits available.

20.7. I2C ISOLATORS

Yes they do exist. For a long time these had to be hand constructed out of separate transistors or gates, a bunch of passive components and some optocouplers.

Both Analog Devices and Silicon Laboratories have fully integrated devices available.

Commonly used I2C devices — LabWorX

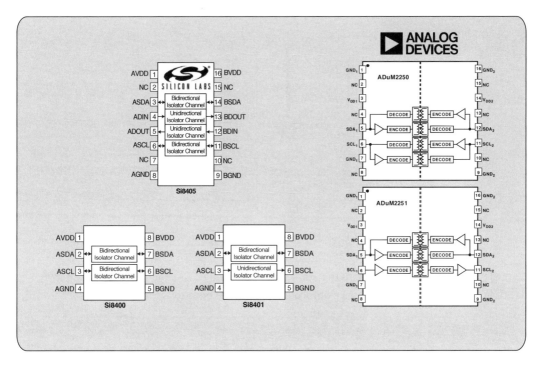

Silicon labs has used their extensive experience in building integrated, galvanic ally isolated, telephone line interfaces to build the I2C isolators family. Depending upon the device selected you can have both SCL and SDA be bidirectional or only SDA be bidirectional.

Analog Devices uses a different approach. They modulate the digital signals onto a carrier and send it across an integrated transformer. Similar to the offerings of Silicon Labs these devices have SDA bidirectional and SCL either unidirectional or bidirectional.

These integrated circuits have two power and two ground pins, as they require power from both sides. The galvanic isolation needs to be maintained between the supplies. If no access is available to one of the source supplies then you can add a small dc/dc converter with galvanic isolation to the system.

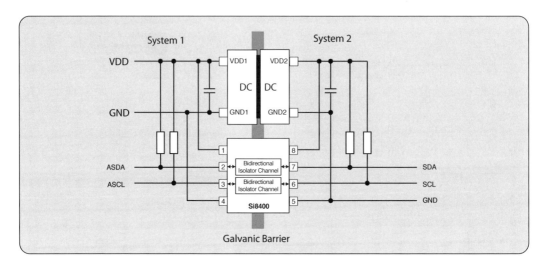

Besides the fully integrated solutions there are other possibilities. One method uses the P82B96 device in conjunction with two passive optocouplers or active optoisolators.

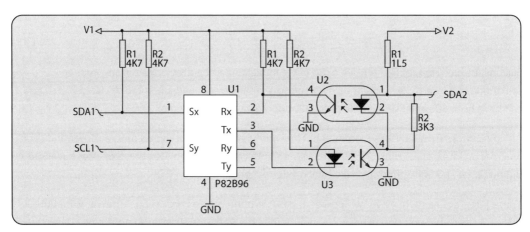

Details have already been given earlier in this book.

Another solution is the LTC4310 made by Linear Technologies. This device can take the I2C bus on one side and pass it across a barrier device such as a transformer or series of capacitors. Depending upon the selected barrier device, large isolation voltages are possible. Since this solution requires two chips you can even create level shifting applications with high common mode differences.

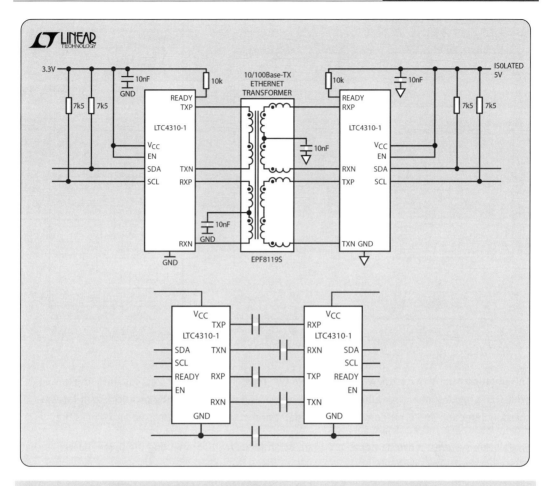

20.8. BUS MULTIPLEXERS AND EXPANDERS

When available address space is exhausted or you simply have too many devices on the bus to meet the electrical requirements, you can insert bus multiplexers. These devices allow you to switch between multiple busses, connecting only one sub bus to the master bus at a time.

20.8.1. Bus Multiplexers

NXP, Linear Technologies, Hendon and others have a whole range of such multiplexers. Under certain conditions, you can cascade such multiplexers to make very deep I2C networks.

Some of these buffers are simple switches but others have active pull up circuitry that can alleviate the effects of capacitive loading on the signal integrity. Other devices offer high voltage level shifting capabilities.

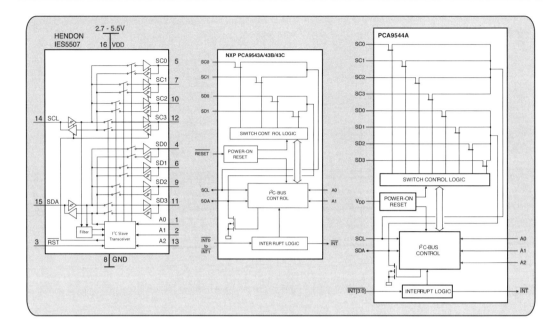

20.8.2. Master multiplexers

The situation can be reversed, what if you have two masters that want to access the same devices? While I2C provides inherent multi master capability it is not fool proof, the protocol itself foresees collision detection and error correction but what if one of the masters locks up and blocks the bus?

For fail safe systems, it can be necessary to unlock the bus by disconnecting the failed master so that a backup system can take over.

This PCA9541A is unique in the sense that it listens to both incoming I2C channels simultaneously. When a command is sent from a master to connect to the downstream (slave) bus the interrupt line of the other master connection is asserted. A master can check the status of the switch by reading a status register.

The device has a state machine that can recover the slave bus should it remain stuck after the crash of one of the masters. The recovery mechanism will send 9 clock pulses with a nACK followed by a STOP. This will free a bus irrespective of if it was in read or write mode and upon completion of this mechanism, the requesting master will receive an interrupt.

Bus switching will only occur when the requesting master has sent a STOP to finish its operation with the PCA9541. Under normal circumstances, the masters must check the status registers to determine whether the bus is free or in use. They can also elect to override this so as to overrule the other channel.

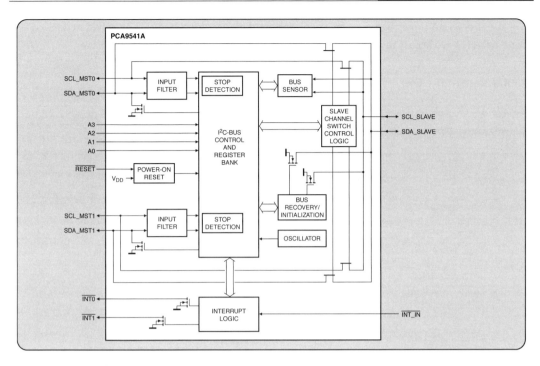

This is a fairly complex device to use correctly. Especially the fail safe mechanism needs careful consideration. The datasheet has detailed instructions and contains examples on how to write the master code to take full advantage of the system.

Note that the masters cannot see each other, so no communication is possible between them. While not connected to the common slave bus, the isolated master can still talk to any other device(s) on its own bus.

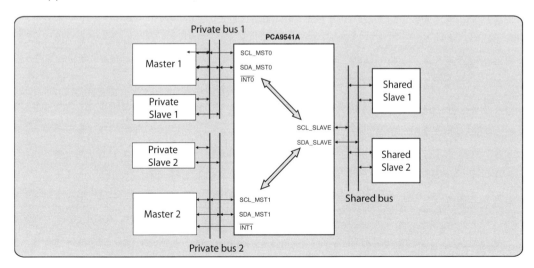

The above diagram shows two masters each sporting their own local bus with attached slaves and both masters have access to a number of shared slaves through the PCA9541 multiplexer.

20.8.3. Roll Your Own

You can also roll your own by using a simple analog bus multiplexer such as a DG409 or 4052. More information is given elsewhere in this book.

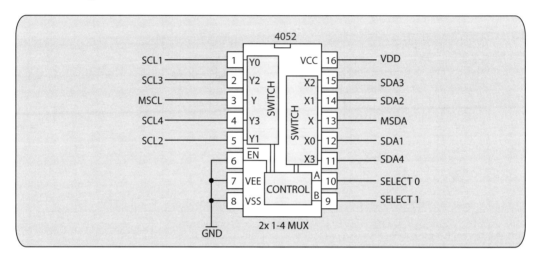

21. LABSTICKS I2C INTRODUCTION

Elektor e LabStick

LabSticks are a group of ready to run mini applications that accompany the LabworX series of books and can be used for prototyping or as building blocks in a system. Each stick can work standalone. The sticks can be plugged onto a breadboard or can be connected to each other. This pluggability is where their name comes from. Think of them as the butter for your bread (board).

This particular book has 2 collections of LabSticks available

Stick Number	Description	Collection
1- 1	I2C Probe	Board 1
1- 2	I2C Stick Power supply	Board 1
1- 3	24xxx EEprom stick	Board 1 & 2
1- 4	PCA9553 PWM LED controller	Board 1
1- 5	PCA9554 LCD / Keyboard module	Board 1
1- 6	LM75 Thermometer module	Board 1 & 2
1- 7	PCF8563 Real time clock with backup	Board 1
1- 8	PCA9554 High current 8-bit output	Board 1 & 2
1- 9	PCA9554 / PCF8574 Protected 8-bit input	Board 1 & 2
1- 10	MCP4725 Dual channel D/A Converter	Board 2
1- 11	ADC121 Three channel A/D Converter	Board 2
1- 12	PCF8591 A/D D/A converter	Board 2
1- 13	MCP40D17 Potentiometer	Board 2
1- 14	PCA9544 I2C Expander	Board 2
1- 15	PCF8574 / PCF8574A / PCA95xx Universal 8-bit I/O stick	Board 2
1- 16	SAA1064 4 Digit LED display (7 Segment)	Board 2

21.1. BOARD 1

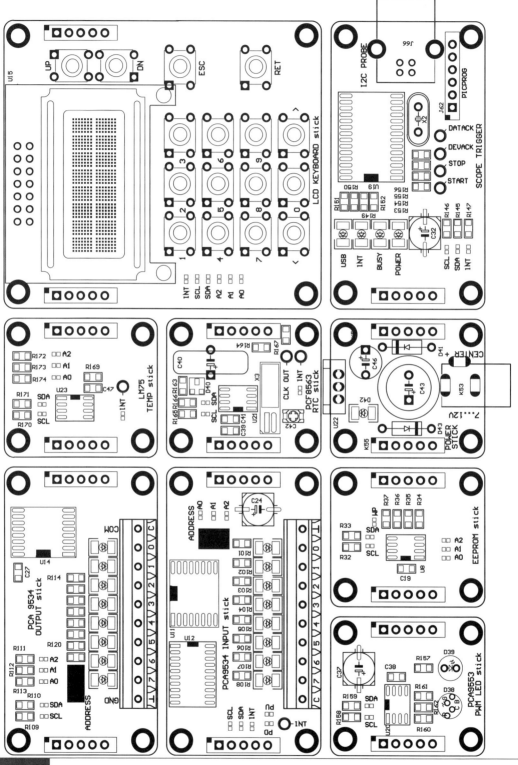

21.2. BOARD 2

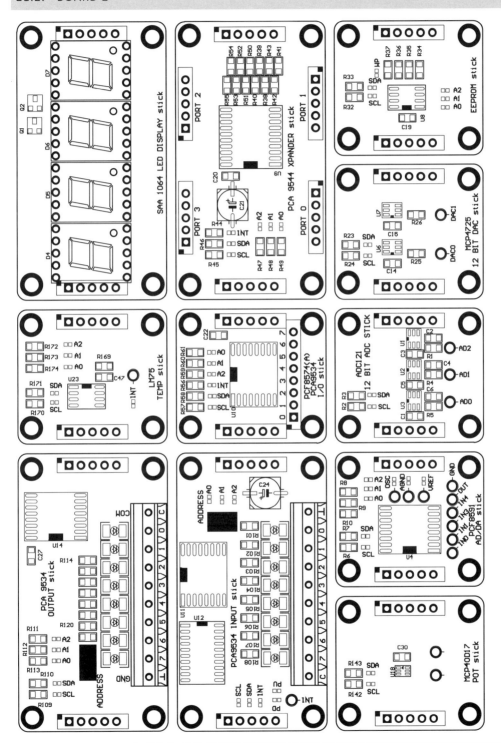

22. LABSTICK 1-1: I2C PROBE

The I2C probe stick is a universal I2C Bus exerciser that allows you to experiment with various I2C devices from a PC environment.

The device is connected to a USB port and is powered from the I2C Bus connector common to the I2C LabSticks. The probe has a number of status indicators and additional outputs to enable the triggering of a scope. The probe implements a human readable protocol to communicate with the I2C bus.

The I2C probe is built around a PIC18F2550 with USB interface. The firmware implements a HID compliant driverless USB interface. The advantage of the driverless mode is that it works with any operating system and no OS specific drivers are required.

The probe will be recognized as a device called 'I2CProbe'.

The probe implements a human readable format parser that can accept command strings up to 32 characters long. Upon execution, the full string is returned with a modified content depending on the commands. Either you can control the probe from your own software using your favourite programming language, or you can use free tools like HIDterminal to talk to the probe.

Go here: http://www.mikroe.com/forum/viewtopic.php?f=88&t=25493 To download a copy.

The LabworX website has sample code to talk to the probe as well as detailed example devices.

22.1. PROBE HARDWARE

The hardware of the probe provides not only the SCL and SDA signals but also shows the user visually when a transaction is occurring or an interrupt is pending. The probe has 4 trigger signals available that activate depending on certain bus conditions. These can be used to trigger an oscilloscope or a logic analyzer to capture transactions.

The I/O pins of the PIC processor are not open drain, as this would pose problems when controlling the actual bus. A trick is used whereby the actual I/O register is always programmed as logic 0 and the code toggles the tristate bit of the I/O driver cell. This has the net effect that the pin is either floating (tri stated), or is actively driven low by the stored 0 when the driver is enabled.

It is the job of the pull up resistors on the bus to raise the SDA and SCL lines back to their high state when needed.

LabStick 1-1: I2C Probe — LabWorX

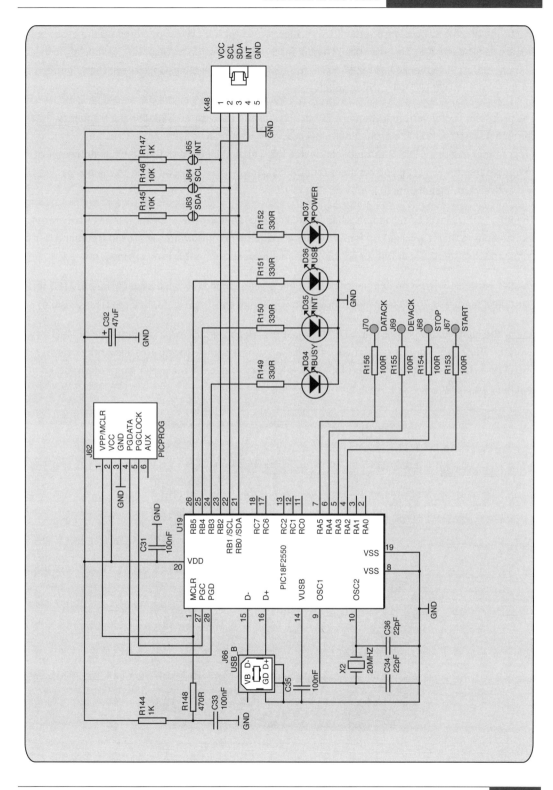

The probe obtains its power from the I2C interface connector K48. When powered up the probe will tristate its SDA and SCL lines and start listening for a USB connection through J66. As long as no valid USB enumeration sequence has taken place, the USB LED D36 will blink. Once connection has been established, the USB LED will turn on solidly.

The processor is driven from a standard 20MHz crystal and derives the USB working frequency through its internal PLL. The probe board has provision to connect a standard Microchip programmer like the PICKIT 2 through connector J62. This can be used to flash a blank processor or to update the firmware. Pull up resistors for SDA, SCL and INT can be connected to the bus by respectively closing jumpers J63, J64 and J65.

The Busy LED D34 will be turned on while an I2C transaction is in progress. The LED has a software monostable so that it will light up for a fixed length of time no matter how short the command is. If commands are then sent at full speed this gives the impression the LED is on permanently.

The INT led D35 will turn on whenever the INT line is being pulled low. Just as with the busy LED this signal also features a software monostable that will extend short pulses so they are clearly visible.

The probe contacts J67, J68, J69 and J70 bring out the trigger signals for an oscilloscope or logic analyzer. During the various state changes of the bus , these signals toggle This allows the user to visually overlay the triggers and therefore makes it easier to decode the actual bus stream should you need to analyze it.

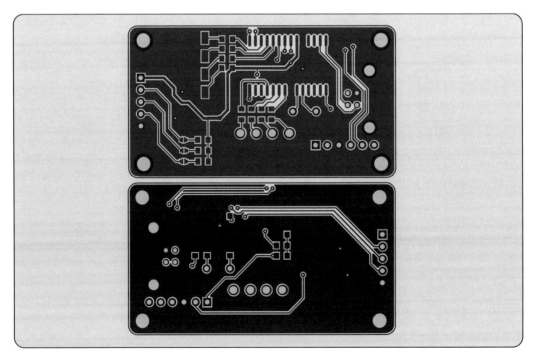

22.2. COMMAND SET

The probe offers command based as well as bare metal access to the I2C bus. The code functionality is master only. The precompiled binary is available for download for people wanting to flash their own probe, or you can order a pre-programmed device if you do not have the means to program the processor yourself.

All commands are in human readable ASCII. The command parser modifies the incoming stream as it is parsing and returns the entire stream to the user.

All streams must be terminated by a <CR> (ASCII code 13) or <CR><LF> pair. The returning stream always terminates on <CR><LF>. Streams can be up to 32 bytes long including the ending <CR><LF>, therefore the maximum useful payload is exactly 30 characters.

Note that the probe is case sensitive! All commands must be written in uppercase. This is done to speed up the processing and minimize the decision tree in the logic. Even though the processor runs fairly fast no time is therefore wasted on case conversion.

All commands, with the exception of the clock speed setting, are single letters. All data is formatted as 2 character hexadecimal (0...9 and A...F). Any command not understood will be replaced by a question mark in the return stream. The probe performs no error checking on the command stream, so it will execute the stream as presented to the best of its abilities.

The parser therefore assumes nothing and will not correct mistakes on your part.

A command stream does not need to be closed, in other words it is not required to 'close' an I2C 'transaction'. You can send commands one character at a time, or send them as a packet. The command parser doesn't care (with the exception of the K command and data, which must always be sent as 2 characters.)

It is important not to send a new command before you have received the return stream, as overlapping command streams will overwrite the internal buffer in the probe and you may corrupt your data. Depending on the operating system the USB handler may implement a double buffered response mechanism. If you cannot read/see the return packet, it will be stored in this buffer. This will cause a de-synchronization, when you do need the return packet for example, when reading from a device.

22.3. BASE COMMANDS

The base command set of the probe allows you to control the bus using simple state commands.

Command	Meaning	Return
S	Start: this sends a START condition on the bus. Upon exit both bus lines will be in a logic low state.	S
P	Stop: this send a STOP event on the bus. On exit the bus lines are tri stated and the bus is no longer occupied.	P
R	Restart: this creates a bus restart event. This is typically used to go from write to read mode.	R
Q	Query ACK: this samples the acknowledge state. If the acknowledge works the letter Q will be replaced with a 'K' in the output stream. If acknowledge fails the output stream will have the Q replaced by a ' '.	K if acknowledge was received. ' ' If no acknowledge was received
G	Give ACK: send an acknowledge from MASTER TO SLAVE. After receiving a byte from a slave this is the kind of acknowledge that must be given.	G
N	Give nACK: this sends a not ACKNOWLEDGE from master to slave. This is used to terminate the reading operation prior to a stop condition	N

22.4. DATA TRANSPORT

Besides these commands there is also data that needs to be moved from and to the bus.

Data	Meaning	Return
00 to FF	A 2 character hexadecimal number representing a byte. This byte is PUT on the bus by the master.	00 to FF
XX	A 2 character placeholder. The probe will perform a bus ready of a byte and overwrite the double X with the hexadecimal notation of the received data.	00 to FF

It is mandatory that data is transmitted as a double character. Should leading zeros be suppressed, the parser will mistake the command for a data character and wrongly calculate the outgoing byte.

It is also mandatory that a data read request is always sent as two characters (XX) as the command parser will overwrite these 2 characters with the return byte encoded as a 2 digit hexadecimal number.

22.5. SIMPLE STREAM EXAMPLES

The following examples show a couple of basic I2C operations as performed using the I2CProbe.

22.5.1. Write a single byte to a slave device

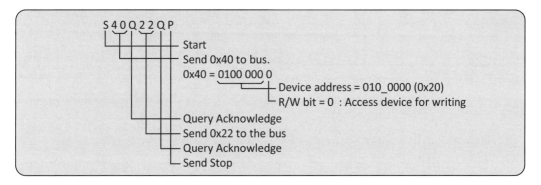

The above datagram sends a single byte to slave device addressed in write mode (R/W = 0).

If the return string reads S40K22KP then all went well. If the return string contains dashes (--) then no acknowledge was received from the slave device.

Note	An Acknowledge Query MUST be sent before a STOP operation. Almost all devices only latch in the received byte upon receiving the Acknowledge clock cycle. I have seen a few software implemented devices that didn't care, but normally an I2C devices does require the Acknowledge cycle before the stop operation. If your communication seemingly doesn't work: check for the presence of the acknowledge cycle.

22.5.2. Write multiple bytes to a slave DEVICE

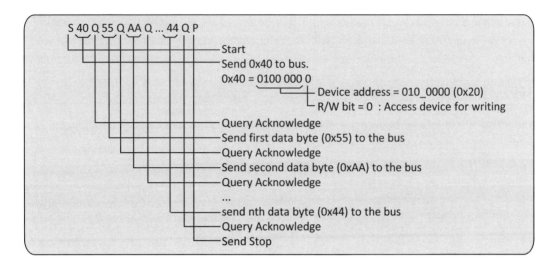

22.5.3. Reading a byte from a slave

This datagram requests a single byte from a device addressed as slave.

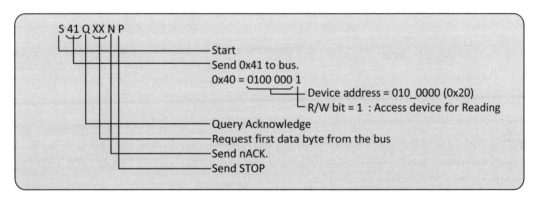

> **Note**
>
> A nACK must be sent before a STOP operation if you want the internal read pointer to increment. It is also possible to directly invoke the STOP operation but this will not increment the internal read pointer in the device.
>
> If you want to read the same byte over and over again, you can simply generate stop directly without the nACK phase.
>
> Not all devices handle this properly so you need to verify this first.

22.5.4. Reading Multiple Bytes from a Slave:

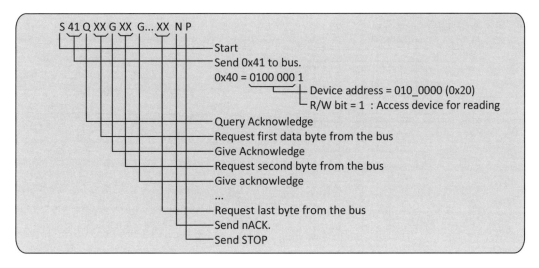

The procedure to read multiple bytes is similar to that of a single byte; the key difference is that the MASTER must give acknowledgments to the slave device by sending a G.

The probe will replace the XX with the received data in the return string.

> **Note**
>
> It is important NOT to send a G but an N prior to the stop command. Should acknowledge be given it may not be possible to create a STOP event on the bus, if the bit transmitted is a zero it will be impossible to raise the SDA line to create the STOP condition.
>
> If you have trouble stopping communication during a read: check for the presence of an acknowledge cycle before the STOP event.

22.5.5. Reading a byte from a Sub address using restart

The probe supports the bus restart operation. This condition is used when switching from write to read mode without giving up control of the bus as with sending a STOP operation. For more details on the restart operation please consult the earlier chapter on restart.

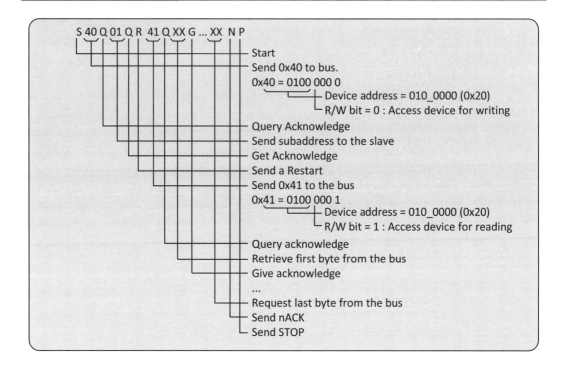

22.6. SUPPORT FUNCTIONS

The probe has additional support commands that allow you to perform various operations.

Command	Description	Return
I	Interrupt test. This tests if the interrupt line is asserted.	'I' if no interrupt is asserted (INT line in high state) '#'if interrupt is asserted (INT is low)
T	Trigger: This generates an interrupt on the interrupt line. The duration is given by the clock setting of the I2C bus. You can use this to test your master code.	T
@	Attention: this forces a reset of the I2C bus. The probe will attempt to clear the bus. It will generate a pattern of clocks and stop operations to attempt to free a stuck bus.	'>' If the bus is clear (SDA and SCL are high) '$'if SDA is stuck low '%'If SCL is stuck low '&'if both SDA and SCL are stuck low
Kx	Setting clock speed where x is 0-9 0 = 1 uS 1 = 10 uS 2 = 25 uS 3 = 50uS 4 = 100 uS 5 = 200 uS 6 = 400 uS (default) 7 = 800 uS 8 = 1600 uS 9 = 3200 uS	This command must be given as one operation, it cannot be split into 2 USB packets.

> **Note**: The K operation must be given as one USB packet or as part of a USB packet. You cannot send the K without a number by itself. The command parser requires the argument. The bus timing intervals are approximate since the I2C command parser can be interrupted by the USB packet engine and be required to service USB interrupts. USB takes highest priority as the computer is unforgiving to unresponsive devices.

22.7. LOW LEVEL FUNCTIONS

The probe has a number of low level features that let you control the SDA and SCL lines directly.

Command	Description	Return
Z	Tri state the bus. This sets both SDA and SCL to tri state. This effectively isolates the probe form the bus. Any other command will automatically pull it out of this mode.	Z
/	Make SCL logic high	/
\	Make SCL logic high	\
-	make SDA high	-
_	Make SDA low	_
#	Sample the state of SDA	1 or 0 depending on the state SDA
+	Sample the state of SCL	H or L depending on the state of SCL

Example:

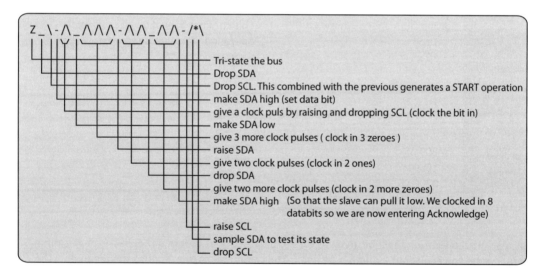

22.8. ASSEMBLY DRAWING

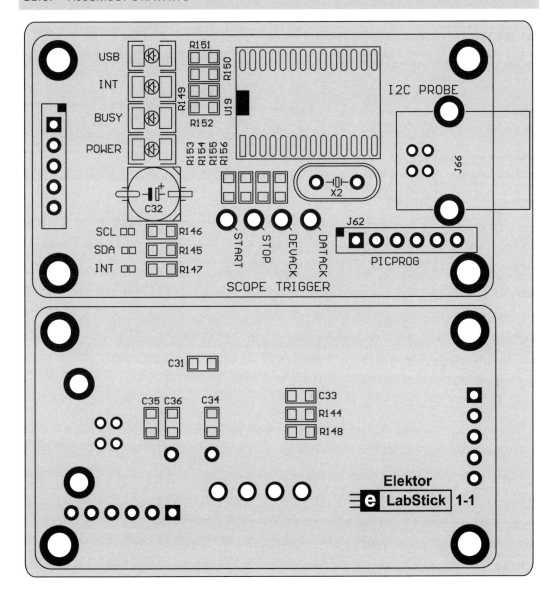

22.9. BILL OF MATERIALS

R153, R154, R155, R156	100 Ohm SMD 0805
R149, R150, R151, R152	330 Ohm SMD 0805
R148	470 Ohm SMD 0805
R144	1K Ohm SMD 0805
R145, R146, R147	10K Ohm SMD 0805
C34, C35	22pF 50V NP0 ceramic capacitor SMD 0805
C31, C33, C35	100nF 25V X7R ceramic capacitor SMD 0805
C32	47uF 16V Electrolytic capacitor
D34, D35, D36, D37	LED PLCC-2 (TOPLED) colour of choice
X2	20 MHz Crystal RJ49/U package
U19	PIC18F2550 with I2CPROBE.BIN loaded
K48	100 mils (2.54mm) pins. Strip 5 pins.
J62	100 mils (2.54mm) pins. Strip 6 pins.
J66	USB connector Right Angle B Type

23. LABSTICK 1-2: UNIVERSAL POWER SUPPLY

The I2C sticks collection needs a small universal power supply that can power a group of sticks. Even though your setup may already have a supply available this design may still come in handy if you have different supply voltages in your system.

The design is built around the classic 7805 (5 volts) but can also be fitted with a 78033 (3.3 volts) or any other 78xx series devices up to a 7812. If you also change the working voltage of the capacitors you can even go beyond that. For the I2C sticks only the 7805 is a suitable part.

The base system is designed to be powered from a small ac/dc adapter or wall-wart. If a heat sink is mounted on the regulator there is ample current available to power a collection of sticks.

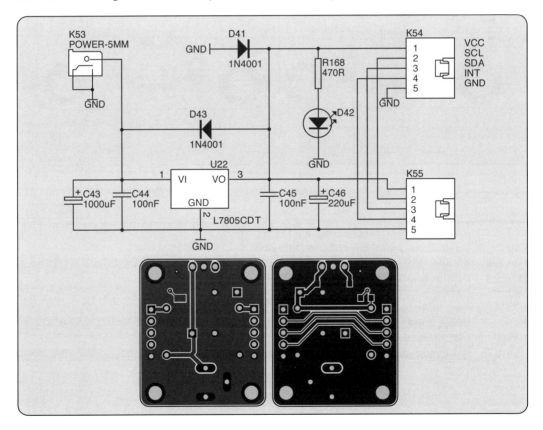

The power supply has the I2C sticks standard connectors on both sides of the board so you can easily daisy chain them if required. The output and the regulator are protected by a parallel (D41) and bypass (D43) diode. A single led (D42) shows that output voltage is available. The value of R168 on the backside of the board will need to be adjusted depending on the installed regulator.

23.1. ASSEMBLY DRAWING

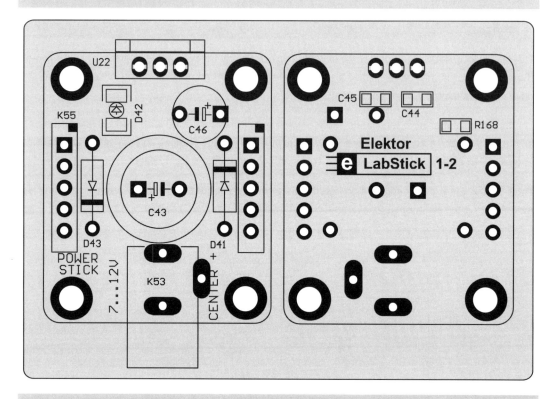

23.2. BILL OF MATERIALS

C43	1000uF 25v radial electrolytic 12mm body
C46	220uF 16v radial electrolytic 7mm body
C44, C45	100nF 25V X7R ceramic capacitor SMD 0805
D42	Led yellow SMD PLCC-2 (TOPLED)
D41, D43	1N400x
U22	See table below
R168	See table below
K53	DC power jack 3.5mm PCB mount
K54, K55	100 mils (2.54mm) pins. Strip of 5 pins.

23.3. SELECTING THE OUTPUT VOLTAGE

This table shows the parts to be changed depending on the desired output voltage of the regulator

Output voltage	Installed regulator (U22)	LED resistor (R168)
3.3 volts	78033	180 ohm
5 volts	7805	330 ohm
9 volts	7809	680 ohm
12 volts	7812	1 K ohm

24. LABSTICK 1-3: 24XXX EEPROM

The EEprom stick allows easy integration of an EEprom device in your design. The board accepts any SO-8 type EEprom such as the 24xx(x) family or PCF8582. Even I2C compatible ram memory can be installed, although these are becoming scarce. Besides the classic EEprom also Ferro electric rams such as the 24Fxx series from Ramtron can be installed. These devices have as additional bonus that they don't have a write cycle.

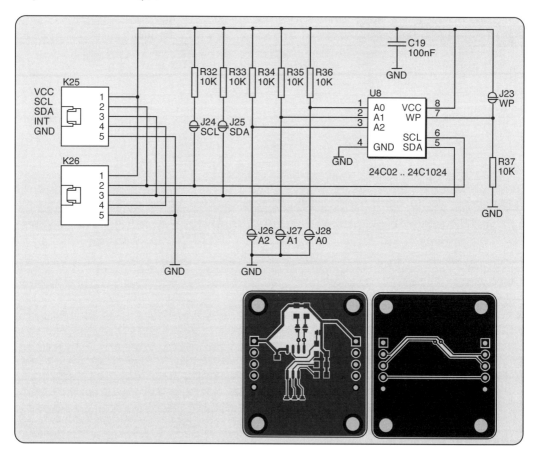

The board allows easy selection of the device address using the solder bridges marked A0 to A2. Closing the bridge sets that address bit to ZERO, opening the bridge sets the address bit to 1.

The SCL and SDA pull up resistors, R32 and R33 can be enabled by closing the solder bridges for SCL and SDA. The EEprom can be write-protected by opening the solder bridge marked WP.

24.1. DEVICES

The table below gives an overview of the most commonly used devices. All these devices will fit the board.

DEVICE	SIZE	BASE ADDRESS	NOTES
PCA8581	128 * 8	1 0 1 0 A2 A1 A0	J63 MUST BE CLOSED
PCF8582	256 * 8	1 0 1 0 A2 A1 A0	J63 MUST BE CLOSED
PCF8594	512 * 8	1 0 1 0 A2 A1 P0	PAGED DEVICE, CLOSE J63
PCF8598	1024 * 8	1 0 1 0 A2 P1 P0	PAGED DEVICE, CLOSE J63
PCF85116	2048 * 8	1 0 1 0 P2 P1 P1	PAGED DEVICES
PCF85102	256 * 8	1 0 1 0 A2 A1 A0	NO WRITE PROTECT
PCF85103	256 * 8	0 0 1 0 A2 A1 A0	
24C01	128 * 8	1 0 1 0 A2 A1 A0	
24C02	256 * 8	1 0 1 0 A2 A1 A0	
24C04	512 * 8	1 0 1 0 A2 A1 P0	PAGED DEVICE
24C08	1024 * 8	1 0 1 0 A2 P1 P0	PAGED DEVICE
24C16	2048 * 8	1 0 1 0 P2 P1 P0	PAGED DEVICE
24C32	4096 * 8	1 0 1 0 A2 A1 A0	
24C64	8192 * 8	1 0 1 0 A2 A1 A0	
24C128	16384 * 8	1 0 1 0 A2 A1 A0	
24C256	32768 * 8	1 0 1 0 A2 A1 A0	
24C512	65536 * 8	1 0 1 0 A2 A1 A0	
24C1024	131072 * 8	1 0 1 0 A2 A1 P0	PAGED DEVICE

I2C compatible EEPROMs can be categorized into 2 main groups: devices with up to 16Kbits and devices with more than 16Kbits.

24.1.1. Up to 16Kbits

The original device was the PCF8581 128x 8 byte EEprom, very quickly followed by the PCF8582 that doubled this amount of memory. Xicor introduced the 24C01 and 24C02 EEProms that are pin and function compatible. Shortly afterwards, other manufacturers like Atmel, ISSI and others followed with their own 24Cxx series devices.

These devices use an 8-bit pointer to allocate bytes in the memory array. This gives a maximum of 256 addressable memory locations. With the introduction of the 24C04 the 8-bits were not sufficient. The solution was to no longer use the A0 pin but to give the device 2 base addresses. One of the device address bits is now used to select a 'page'. The 24C04 device listens to 1010xx0 and 1010xx1. The combinations for xx are set by pins A2 and A1 (pin A0 is not used on the device. The device will respond to being addressed with bit a0 set to 0 and to 1)

If the device is being addressed at 1010xx0 the bottom half of the memory is accessed. If the device is accessed as 1010xx1 the top half is accessed. This is called paging.

You can think of this paging mechanism in another way: as having a board with multiple 24C02's.

Let's assume you have a board with two 24C02 or PCF8582 devices. The first one has as base address 1010000 while the second one sits at 1010001. If you look at the addresses of the devices you see that in this case it is the LSB that determines which chip is being talked to. This is exactly what a 24C04 does. It will respond to both addresses and only use bit A0 to switch internal pages.

The 24C08 emulates in this fashion four 24C02 and uses bits A1 and A2 of the device address to select which page is accessed. The 24C16 uses all three bits.

This means that you can install only one 24C16 on an I2C system, just as you can install only two 24C08 devices or four 24C04 devices.

24.1.2. Above 16Kbits

Devices with more than 16Kibits (2048 *8) had run out of possible address space. With the introduction of the 24C32 the internal address pointer has been changed from 8-bit to 16-bit. This now allows for a maximum of 65536 * 8 or 512 Kibits (Kibits or Kibibits are now official IEEE and IEC notations for binary numbers. The traditional Kilobit means 1000 bits while the KiBit or KibiBit means 1024 bits. There's lots of debate about it and uptake is slow but these names have now been ratified into an official standard.)

Since the internal address pointer is a 16-bit number there are no more restrictions on the usable I2C addresses. There is one exception though: for devices larger than 65535 *8 this restriction is reemployed. The 24C1024 uses A0 bit of the I2C bus to select a page. This device behaves as two 24C512s on consecutive I2C addresses. At the time of writing this is the largest I2C device out there. There was brief mention somewhere that someone actually made a 24C2048 but apparently it never went into production.

24.2. USAGE

Since the three groups of devices behave differently, I will split this chapter into the appropriate subtopics.

24.2.1. Devices up to and including 24C16 (8-bit memory address)

The device is accessed in write mode by sending the device address followed by the EEprom target address as a single byte. Subsequent bytes represent the data to be written. Care should be taken not to overrun the page end. More about the paging mechanism will be given later on in this book.

Byte	Direction	D7	D6	D5	D4	D3	D2	D1	D0	CONTENT
0	OUT	0	1	0	0	A2	A1	A0	1	ADDRESS
1	OUT	P7	P6	P5	P4	P3	P2	P1	P0	MEMORY POINTER
2..n	OUT	D7	D6	D5	D4	D3	D2	D1	D0	DATA

For the 24C01 only, an address from 0 to 127 is valid. If memory cell 128 is accessed, the physical location will actually be location 0, cell 129 returns the contents of cell 1 and so on.

For devices larger than 256 * 8 the complete memory pointer is formed by P7 to P0 and one or more bits of the address field. These devices react to more than one I2C address.

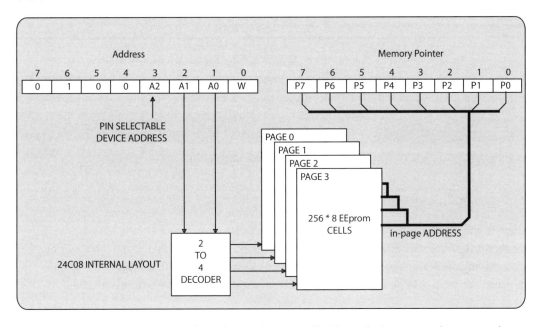

The above image gives you an idea how this works internally. These devices are nothing more than a collection of 24C02's located on consecutive I2C addresses.

Remember that the address counter will roll over when it hits a page boundary in write mode! Later on I will go into more detail on this behaviour.

24.2.1.1. READ MODE

When accessed for read the device will continue sending out data as long as reads are performed. The address counter will auto increment, for each consecutive byte read. To load a random memory address the device must be accessed in write mode and the 8 byte base address must be transmitted. You can then send a STOP and readdress the device in read mode, or you can simply issue a repeat start and access the device as read again. The EEprom remembers the selected address.

24.2.2. Devices from 24C32 upwards (16-bit memory address)

Once devices became larger than 2 Kilobytes the available address space is exhausted. For larger devices, the address pointer has been changed to be a 16-bit pointer. This makes these devices software incompatible with their smaller counterparts.

Byte	Direction	D7	D6	D5	D4	D3	D2	D1	D0	CONTENT
0	OUT	0	1	0	0	A2	A1	A0	1	ADDRESS
1	OUT	P15	P14	P13	P12	P11	P10	P9	P8	MEMORY POINTER
2	OUT	P7	P6	P5	P4	P3	P2	P1	P0	MEMORY POINTER
3..n	OUT	D7	D6	D5	D4	D3	D2	D1	D0	DATA

Apart from the larger memory pointer these devices behave identically to their smaller counterparts. Due to the larger memory array the page size on these devices has been increased in size.

24.3. PITFALLS WITH E2PROMS

There are a couple of pitfalls with these E2Prom devices that deserve some further explanation.

24.3.1. Operating voltage and voltage range

There is NO consensus amongst manufacturers about the value of the operating voltage or voltage range. A 24C02 from manufacturer X may work between 2.7 and 5 volts while the same part number from manufacturer Y only works at 5 volts, therefore always check the datasheet of the actual device at hand. Do not simply grab a datasheet from a similar part made by someone else.

Not only is there a difference in working voltage, there is also the possible problem of the voltage range. If a device is specified for 1.8 to 5 volts operating voltage, this can mean different things. Sure, you will be able to read it using a supply voltage that falls within range, but you may not be able to write at the low end of the voltage spectrum! You need to carefully read the datasheet of your particular device. Again, make sure you have the exact part number and manufacturer's datasheet. Do not simply grab anything 'similar' from a different manufacturer.

24.3.2. Page writing and page sizes.

All of the EEprom devices support so called page writes. The idea behind it is to speed up individual cell writes.

Problem 1: For every write operation the cell needs to be erased. The EEPROMs are not smart enough to erase only those bits that need to change state. This erasing mechanism erases multiple bits in a block, or a so called 'page'. When you perform a random memory write the EEprom will copy the page contents into a ram, modify that page ram with the contents received and then perform an erase. When the erase completes, the EEprom then stores the data contained in the page ram into the EEprom cells.

This has consequences. The physical write operation only occurs when you send a STOP command. If you write only one byte at a time the device is performing a lot of needless erase operations, therefore you need to bundle your bytes so that they fall inside a page.

Problem 2: A second consequence is that every erase and program cycle causes irreparable damage to the EEprom cell. If you always write single bytes on a 16 byte per page device you will wear it out 16 times faster! This damage occurs due to charge injection into the non conductive material around the floating gate of the EEprom cell. If there is enough charge trapped there the cell will be permanently programmed to logic 1 and cannot be set (written) to a logic low anymore.

Actually, erasing sets a cell to logic 1. It is the programming cycle that sets it to logic low by injecting a charge on the floating gate that puts the transistor in conduction.

Problem 3: When accessing a device with random locations you can still send multiple bytes but you must take care not to roll over the page border.

An example:

Let's consider the case of a device where the page is 4 bytes long. If I write to any random location I am faced with the default program cycle. If I write 2 random locations within the same page this EEprom write time is multiplied by the number of accesses, in this case 4.

So I can speed up writing by putting bytes in order and staying within the page.

Page Address	0				1			
	0	1	2	3	4	5	6	7

Write Address 2, 0x55

		55					

Write Address 6, 0xAA, 0x2C

		55				AA	2C

Write Address 1, 0x11, 0x22, 0x33, 0x44

	44	11	22	33			AA	2C

page rollover

The problem exists if you will only need to update a few consecutive bytes inside a page. Let's consider the following case: I will write 1 byte into the device at address 2.

Address 2 falls within the first page (for a page size of 4 the first page sits at address 0 to 3). If I access the device and send start address 2 followed by 1 byte and then a stop, the device will store this into the EEprom array.

Now, let's write 2 consecutive bytes starting at address 6 which, for this device, falls in page 1. The device will accept the I2C transfer and perform the write as anticipated.

Let's say I want to write 4 bytes starting at address 1. Since address 1 falls within page 0 of the device this is the page where the operation will be performed.

Here is the problem: the page counter inside the EEprom does NOT increment. There is only enough ram to perform operations on one page. If I send 4 bytes the address pointer for the page will roll over and the 4th byte I send will land on address 0!

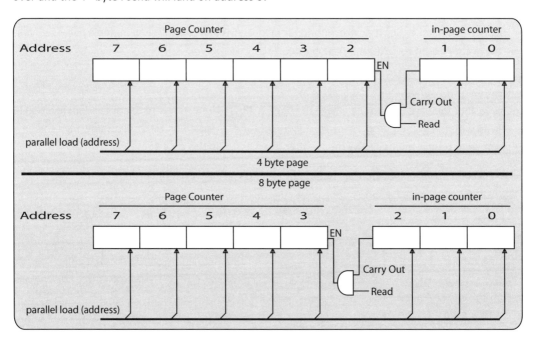

The reason is that internally the EEprom has 2 counters. When you load an address the full counters are preset with the received address. Each consecutive byte received increments only the byte counter and not the page counter. If I were to send 19 bytes into an 8-bit page only the last 8 would be stored. Since the byte counter has overrun multiple times the data will be shifted internally!

There is again no consensus between manufacturers on page sizes. A 24C02 from manufacturer x may have a page size of 4 while the same part from a different vendor may have a page size of 8 or 16 (pages are always a power of two).

24.3.3. Summary

- Optimize your data into pages to decrease the total write time and increase the lifetime of the EEprom.
- Take care not to roll over the page end as this will corrupt data at the beginning of the same page.
- Always check the correct datasheet of the exact manufacturer. There is NO consensus for working voltage, voltage range or page size. You NEED to verify.

24.4. ASSEMBLY DRAWING

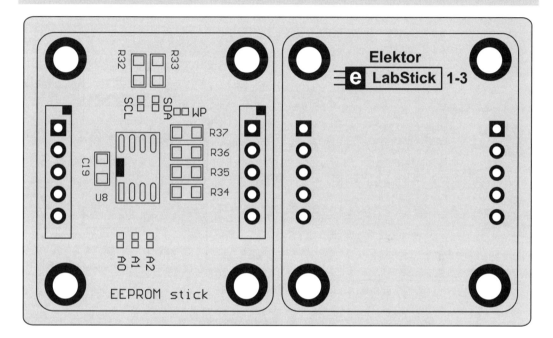

24.5. BILL OF MATERIALS

R32, R33, R34, R35, R36, R37	10K ohm SMD 0805
C19	100nF 25V X7R ceramic capacitor SMD 0805
U8	2401 ... 24512 / PCF 8581, PCF8582 I2C EEPROM
K25, K26	100 mils (2.54mm) pins. Strip of 5 pins.

25. LABSTICK 1-4: PWM LED CONTROLLER

The PWM controller is specifically made for controlling LEDs, but it could be used for applications as well. The PCA9533 or PCA9553 has 4 independent channels that have an open drain output.

Each channel has two programmable repeat rates and duty cycle settings. A control register allows selecting one of four states per output: OFF, ON, Setting1 and Setting2.

The schematic is fairly straightforward and I opted to use one channel for a single led and combine the three other channels to drive an RGB LED.

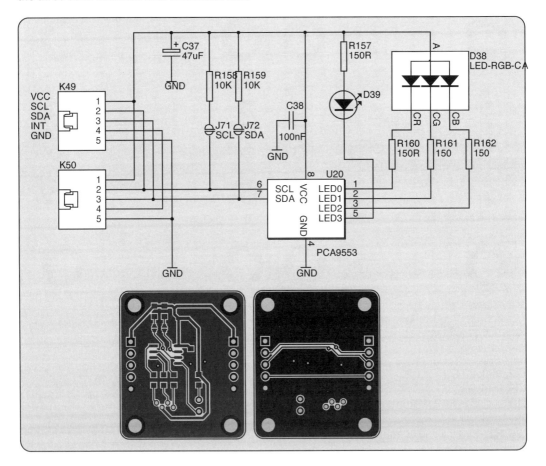

Connectors K49 and K50 have the usual I2C bus interface. Jumpers J158 and J159 allow the enabling of SCL and SDA pull up resistors.

Resistors R157, R160, R161 and R162 limit the current per LED. Depending on the LED used these values may need changing. You need to look up the maximum current allowed through the LED and then calculate the resistor values.

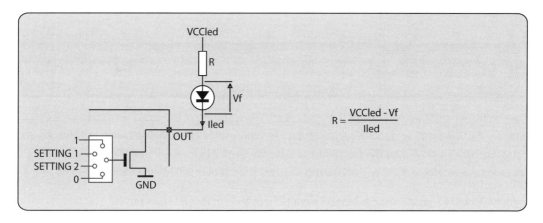

The image above gives the basic equations to do just that. Technically there is still the drop across the output transistor in the chip but this is negligible. The parameters for Iled and Vf can be found in the datasheet of the LED you are going to use.

25.1. ADDRESS

The PCA9533 comes in two versions that feature a different base address. The device itself has no address selection pins. The correct address is set by ordering the part number with the correct suffix.

	MSB							LSB
PCA9533/PCA9553 -01	1	1	0	0	0	1	0	R/W
PCA9533/PCA9553 -02	1	1	0	0	0	1	1	R/W

25.2. USAGE

The PCA9533 has 6 internal registers that are selected using a control register. The diagram below shows the basic register contents of the device. Keep in mind that only one output channel is drawn.

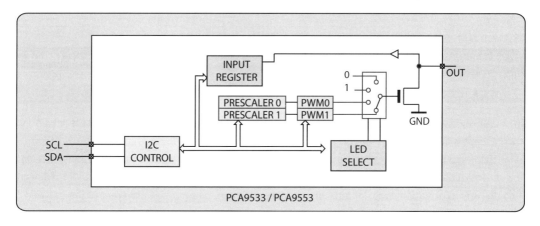

Control Byte	MSB 0	0	0	AI	0	R2	R1	LSB R0

The bits R2, R1 and R0 select one of the internal registers. The AI bit allows an auto increment to happen. If you want to write to all registers you can simply write 00010000 followed by 8 more bytes. Each byte will land on the next address location (the value for R2, R1 and R0 will increment for each byte received). If more than 8 bytes are sent the counter will roll over automatically.

This allows you to update the contents in a continuous stream for all registers.

If you only want to update one register very fast you can turn the auto increment off by setting the AI bit to zero. You can still send a stream of bytes but they will all land on the same internal register that was selected using bits R2, R1 and R0.

A command byte of 0000_0010 would select register 2 and send a stream of bytes in there. This is useful if you only want to update one parameter very fast.

25.2.1. Available registers

Register	R2	R1	R0	Description
INPUT	0	0	0	Read only register that reflects the current output state
PSC0	0	0	1	Prescaler 0
PWM0	0	1	0	PWM register 0
PSC1	0	1	1	Prescaler 1
PWM1	1	0	0	PWM 1
LS0	1	0	1	LED selector

25.2.2. Input register

	MSB							LSB
INPUT REGISTER	0	0	0	0	LED3	LED2	LED1	LED0

The input register reflects the current state of the output pins. Even though this register is read only, it will accept write operations and acknowledge them, but it will not do anything with the write data. The reason behind this is simple: if you put the device into auto increment mode you can send long chains of data to the device. Even if you write to this input register there is no problem. You can send dummy information, wait for an acknowledge, and then continue with the rest of the payload in the transport stream.

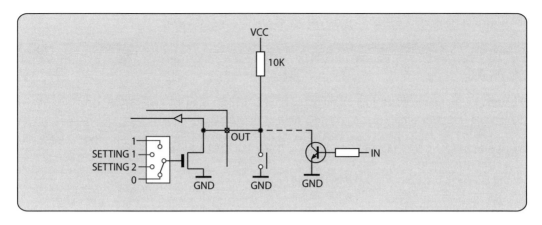

The input register can also be used for other purposes. If the output pin is set to OFF, then the pin can be used as an input by tying it high using a resistor. If a switch, or transistor, pulls the signal low, this can be detected when the register is read. Since the PCA9553 has no interrupt capability this requires polling. For frequently used buttons this may not be a good idea, but it can be an ideal candidate for power control.

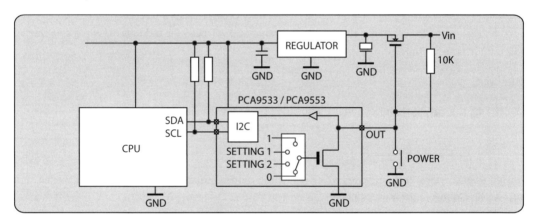

Whenever the button is pressed, the gate of the P MOS is pulled low. This turns on the MOSFET and provides Vin to the regulator, the CPU comes to life and the first thing that is done is to write the output pin to logic low (LED ON). The output pin of the PCA9553 now takes over for the power button. When it is time for the system to go back to sleep the processor writes the output back to 1 (LED off). This brings the P MOS out of conduction effectively shutting off power to the entire system.

You could even get rid of the regulator altogether. If you know the load formed by your system you can program a PWM rate that opens and closes the MOSFET to form a simple switching regulator that keeps the voltage across the electrolytic capacitor constant.

25.2.3. PCS0 and PCS1 register

The prescaler registers are used to program the period of the LED signals.

PCSx	MSB							LSB
	D7	D6	D5	D4	D3	D2	D1	D0

The period is programmed as an 8-bit number. The PCA9533 has an internal oscillator running at 152 Hz. The prescaler simply places a divider on that master oscillator.

The output period is determined as the value of (PSCx +1) / 152 for the PCA9533

For the PCA9553 this equation is (PSCx +1) / 44

25.2.4. PWM0 and PWM1 register

The Pulse Width Modulation registers are used to program the duty cycle of the LED signals.

PWMx	MSB							LSB
	D7	D6	D5	D4	D3	D2	D1	D0

The period is programmed as an 8-bit number. The duty cycle is set as a fraction of the selected period. A value of 128 will give a 50% on time. A value of 0 will turn the led fully off while a value of 255 will turn the led fully on.

Duty cycle = value/256

PWM0 sets the duty cycle for channel 1 while PWM1 set the duty cycle of channel 2.

25.2.5. LED selector register

The led selector register determines which operation mode is selected for a particular LED. Each LED has two bits in this register as a selector.

PCSx	MSB							LSB
	LD3_1	LD3_0	LD2_1	LD2_0	LD1_1	LD1_0	LD0_1	LD0_0

For any given led the 2-bits select the mode of operation.

- 00: LED off
- 01: LED on
- 10: LED connected to PWM0 and Prescaler 0
- 11: LED connected to PWM1 and Prescaler 1

At power-up time all LED's are set to the OFF state.

25.2.6. USING as input

It is possible to use one or more of the LED drivers as an input. Simply program the LED output into the off state by setting its control bits in the LED selector register to 00.

Add a pull up resistor to the output pin and connect a pushbutton or transistor from the output to ground. The state of the pin can now be read from the LED register.

25.3. ASSEMBLY DRAWING

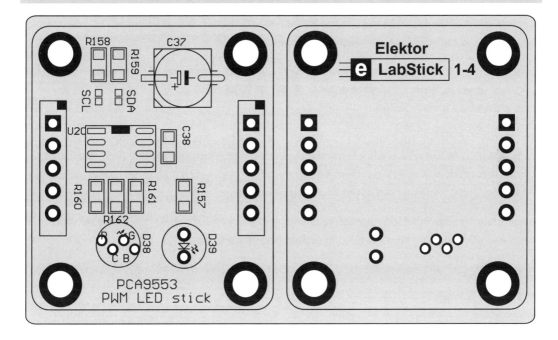

25.4. BILL OF MATERIALS

R158, R159	10 K Ohm SMD 0805
R157, R160, R161, R162	150 Ohm SMD 0805
C38	100nF 25V X7R ceramic capacitor SMD 0805
C37	47uF 16V SMD
U20	PCA9533 or PCA9553
D39	LED WHITE 5MM
D38	LED RGB 5MM Common ANODE
K49, K50	100 mils (2.54mm) pins. Strip 5 pins.

26. LABSTICK 1-5: LCD / KEYBOARD USER INTERFACE

The user interface stick is an elaborate circuit that has an LCD display and a 16 key matrix keyboard. The I/O is handled by the PCA9554 (this was explained previously) in the universal I/O stick. The bus master is responsible for scanning the keypad and handling possible de-bounce/repeat operation.

A standard HD44780 based LCD display is wired up in 4-bit mode to the I/O controller.

Outputs 0 to 4 of the PCA9554 are used as a 4-bit bus to drive data into the display. The lower three bits are sent to the 74138 three-to-eight decoder as well.

In response to the value written, the 74138 will pull one of its outputs low. Each output drives two switches. Series resistors protect the outputs of the 74138 against inadvertent shorts if two or more keys are pressed simultaneously.

The switches feed into two common return lines. These 'sense' return lines go through resistors R130 and R132 to a small capacitor used for debouncing. The resistors R131 and R133 pull up these two wires when idle. If no key is pressed both signals remain at a logic high. When a scan operation finds a key press, it will pull one of the two sense lines low.

By programming the PCA9554 to generate an interrupt on each state change for these two pins (I/O6 and I/O7) you can interrupt the process and handle the key press.

Alternatively you can also simply read back the I/O state and crunch the information in polled mode.

The display operates in 4-bit mode and is put into permanent write mode. The Register Select line of the display is controlled by I/O 4, while the enable pin is used as the write gate and drive from I/O 5.

Resistors R122 and R129 create the bias voltage for the display and set the contrast.

The classical circuitry that provides address selection and allows the enabling of pull up resistors for SDA, SCL and INT is also included on the board.

LabStick 1-5: LCD / Keyboard user interface — LabWorX

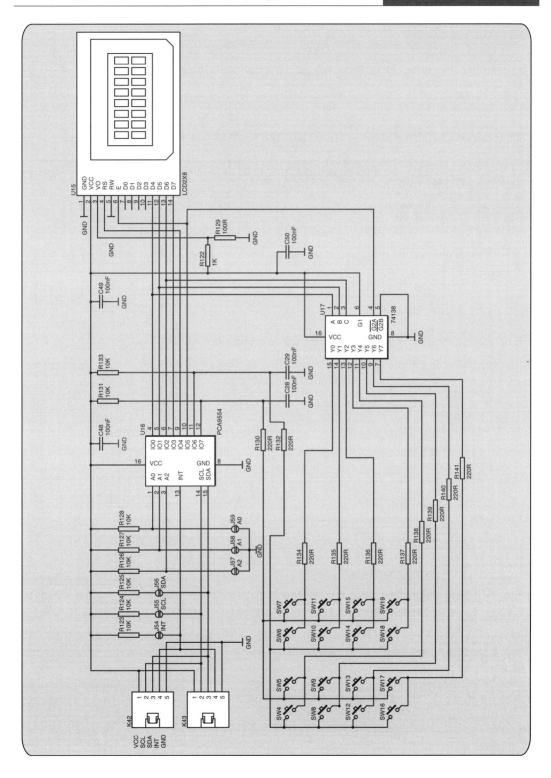

I2C Bus

26.1. INITIALIZATION

This board requires some initialization code before it can do anything.

26.1.1. Device initialization

For this system to work it is necessary to initialize the PCA9554 with the correct settings.

Register	MSB - LSB	Description
CONFIGURATION (R3)	0011_1111	Configure input or output direction
POLARITY (R2)	1100_0000	Key return lines are active low.
OUTPUT (R1)	1111_1111	All signals High.

The PCA9554 is now loaded with the correct settings and we can proceed to the initialization of the LCD Display.

26.1.2. LCD Display operation

The default initialization in 4-bit mode for the LCD is pretty much cut and dried.

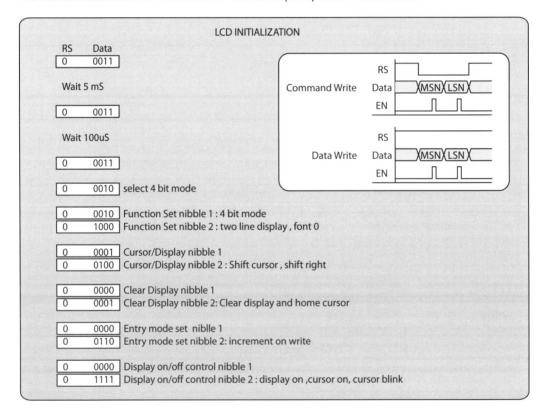

Transfers to the LCD are sent nibble by nibble. The Most significant nibble (MSN) needs to be sent first, followed by the least significant nibble (LSN). Once the initialization is complete, operation is very simple.

The table below shows the command codes for the HD44780 display controller.

Code	RS	R/W	DB7	DB6	DB5	DB4	DB3	DB2	DB1	DB0	Description
Clear Display	0	0	0	0	0	0	0	0	0	1	Clears display and returns cursor to home (Address 0).
Cursor At Home	0	0	0	0	0	0	0	0	1	*	Returns cursor to home and return display to zero position
Entry Mode Set	0	0	0	0	0	0	0	1	I/D	S	Sets the cursor and display move (I/D) direction (S)
Display On/Off Control	0	0	0	0	0	0	1	D	C	B	Turn on or off the display (D), cursor (C), and cursor blink (B).
Cursor/ Display Shift	0	0	0	0	0	1	S/C	R/L	*	*	Moves cursor and shifts display without changing DDRAM contents.
Function Set	0	0	0	0	1	DL	N	F	*	*	Sets interface data length (DL), number of display lines (N) and font (F).
CGRAM Address Set	0	0	0	1	Character Generator RAM ADDRESS					Sets the CGRAM, data is sent and received after this setting.
DDRAM Address Set	0	0	1	Display Data RAM ADDRESS						Sets the DDRAM, data is sent and received after this setting.
CGRAM /DDRAM Data Write	1	0	WRITE DATA							Writes data into DDRAM or CGRAM.

Data transport to the display is simple and straightforward.

26.2. KEYBOARD ACCESS

The keyboard scanning occurs under the control of software on the master CPU. There are two possible ways to do it, one is interrupt driven, the other works in a polled mode.

26.2.1. Interrupt driven mode.

Periodically the master processor will perform a scan sweep of the keyboard matrix. The processor will send a 3-bit number from 0 to 7 on bits D0 to D3 of the I/O port. By keeping the EN signal of the display low this does not change in any form the display portion of this block.

If no key is pressed there will be no change of state on pins D6 or D7. The capacitors C28 and C29 in combination with R131 and R133 keep these at logic high. Since the PCA9554 has built in pull up resistors you could actually play with the capacitor value so you may possibly get rid of R131 and R133 altogether.

Should a key be pressed when the scan is being performed, then one of the sense lines (the lines coming from the capacitors are the sense lines) will be pulled low by the 74138 output.

This state change will cause the PCA9554 to generate an interrupt. The interrupt handler only needs to retrieve the current value of the scan counter (the scan counter is a shared variable) and perform a read operation to retrieve which sense line has been toggled low.

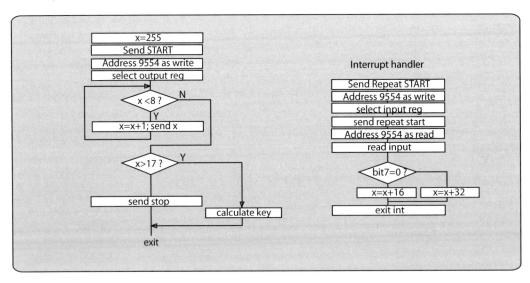

The drawback is that the interrupt handler must abort the pending I2C transaction, send a stop, address the device in write mode to select the correct register, and then access the device in read mode.

The flowchart above gives a general idea. The scan handler (left flowchart) is invoked by a periodic event. After placing a START on the bus the PCA9554 is addressed in write mode and the output register is selected.

A loop now writes in sequence the numbers 0 to 7 to the output register. This loop exits at the end and two extra tests are performed.

The idea is simple: select a scan line and if no interrupt occurs: proceed to the next one. If the interrupt handler kicks in, it switches the PCA9554 back into read mode, selects the input register, gives a repeat start and reads the input register. Depending upon if bit7 or bit6 is low the interrupt handler modifies the contents of X and then exits

At this point we re-enter the scan routine (coming back from the interrupt). If an interrupt has occurred the test for x<8 will now fail. The contents of X now represent a unique number for the key that was pressed.

26.2.1.1. OPTIMISATION

There are various ways to optimize this code for speed and performance.

To really optimize the system you may want to change the scan logic using two 74138 and cross link the gate signals.

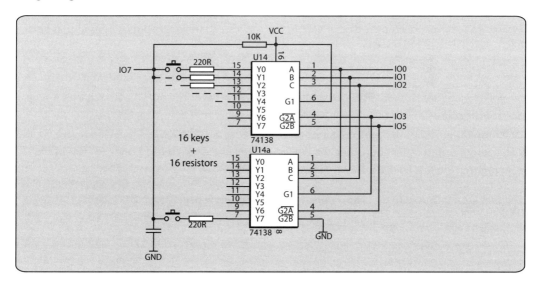

The schematic above shows you exactly how to do this. The decoding is now complete for a 16 key system. In this case the 4-bit code placed on the bus has a direct relationship to the key being pressed. When an interrupt occurs, you have found the key! This method requires no reading of the PCA9554 to find out which sense line was activated.

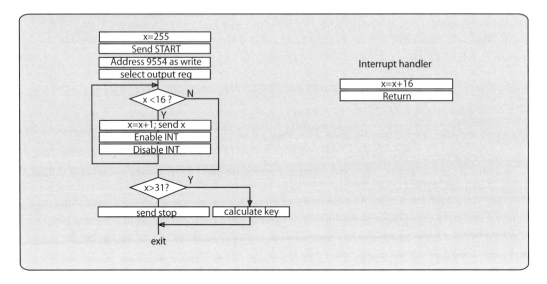

When an interrupt occurs, all we do is to modify the contents of X.

The handler is slightly modified to make sure that interrupts can only be trapped in a small time window. This completely eliminates any false key hits. At start-up I also wait for the interrupt to be clear, as long as the user keeps pressing the button, the routine will wait. This acts as simple debounce logic.

If you want to have a key repeat type of action, you can instead use a timeout counter at that point.

26.2.2. Polled mode

The polled mode is a bit more elaborate but it's still pretty straightforward. The flowchart is a simple loop construction that addresses the PCA9554 in write, selects the output register, writes the contents of X, sends a restart and readdresses it in write mode. Now we select the input register and send another restart. After addressing the PCF9554 in read mode we retrieve the contents of the input register. If bits D7 or D6 are low we have a key push.

If neither D7 nor D6 is zero we send a repeat start and the cycle begins again.

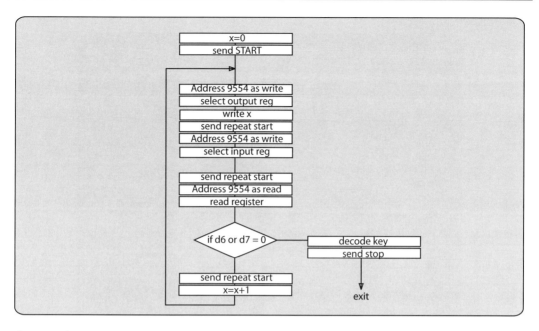

If you don't want to be penalized because of the registers in the PCA9554 you can always swap it out for a PCF8574. In that case the flowchart becomes even easier. Note that a PCF8574 may not work properly on this board given the fact that there are no pull up resistors present on the digital output lines.

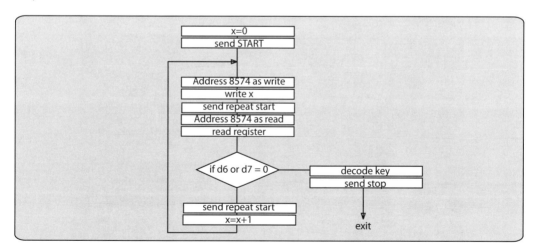

26.3. CIRCUIT BOARD

The next pages show the top and bottom copper of the circuit board and the assembly plan.

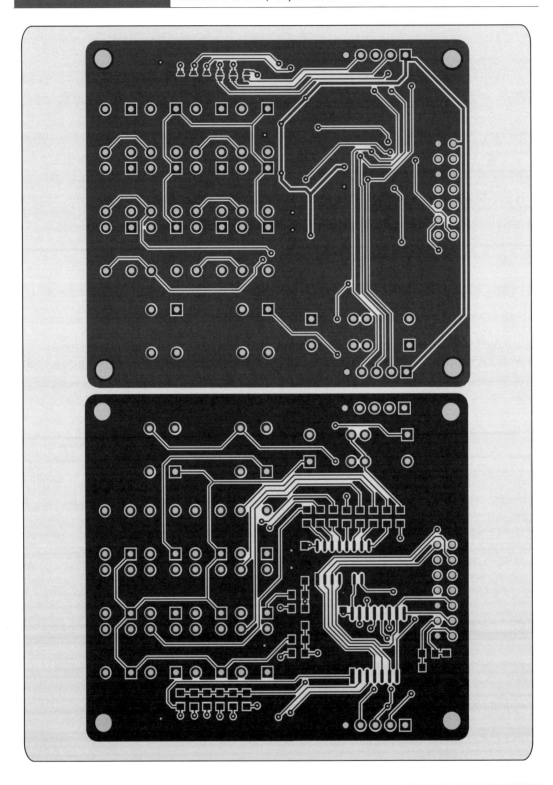

LabStick 1-5: LCD / Keyboard user interface — LabWorX

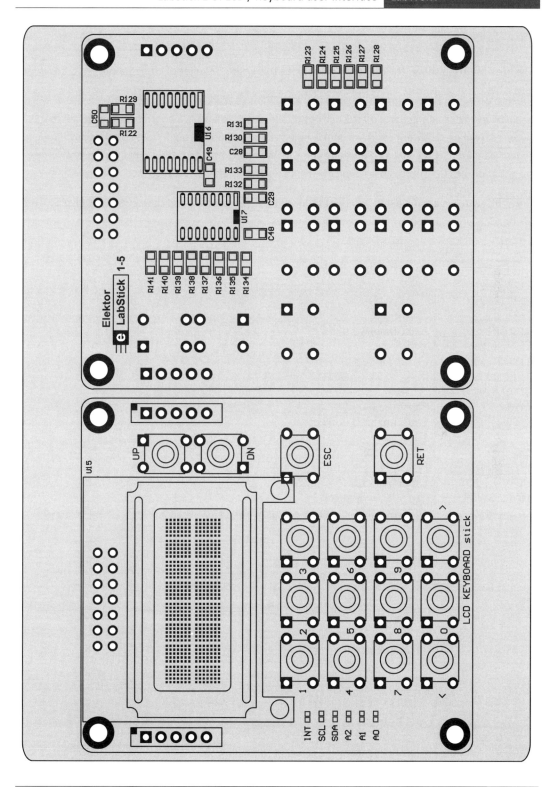

26.4. ASSEMBLY

The previous page shows the assembly plan. Since this board has a lot of components on it, it is advisable to follow a strict order of assembly.

Begin by installing all the parts on the back of the board, then turn the board over and install the switches. I did explicitly opt for the through-hole versions. These through-hole switches have a square plunger that accepts a coloured round cap. When assembled the total height is slightly more than that of the LCD display so it fits nicely through a front panel if you wish.

Once the switches are installed, you can mount the two pin headers and then install the display. The pin headers are installed on the back of the board.

R123, R124, R125, R126, R127, R128, R131, R133	10 K Ohm SMD 0805
R130, R132, R134, R135, R136, R137, R138, R139, R140, R141	220 Ohm SMD 0805
R122	1K Ohm SMD 0805
R129	100 Ohm SMD 0805
C48, C49, C50, C28, C29	100nF 25V X7R ceramic capacitor SMD 0805
U16	PCA9554 or PCF8574
U17	74HCT138 SO-16 PACKAGE
SW4 ... SW19	6x6 MM pushbutton.
K42, K43	100 mils (2.54mm) pins. Strip of 5 pins.

27. LABSTICK 1-6: LM75 (A) TEMPERATURE SENSOR

This stick implements the basic circuitry around an LM75 temperature sensor. This device was originally developed by National Semiconductors for usage on early Pentium class motherboards. The device monitored the temperature of the north and south bridges. A system supervisor could then flag an over temperature condition and the CPU clock speed could be reduced to decrease the heat dissipation from the system.

The LM75 supports 8 sub addresses, allowing for up to 8 devices on the bus. Since its introduction a new variant has been made in the form of the LM75A. Where the original device only had a 9-bit converter, the LM75A variant has an 11-bit converter yielding a much higher resolution.

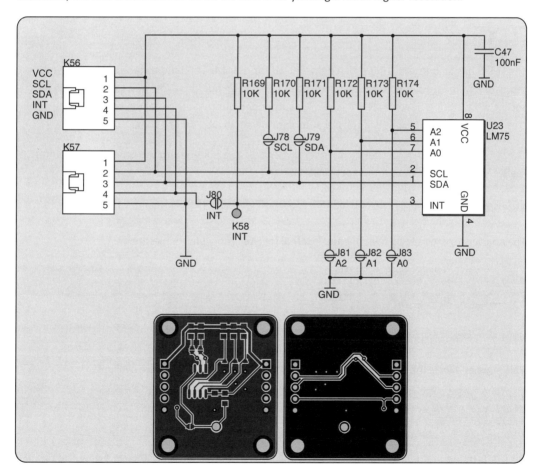

This schematic is another variation on the classic circuitry. J170 and J171 allow you to switch on the pull up resistors for SCL and SDA, J80 connects the interrupt output to the interrupt signal of the interface connector. The LM75 has programmable thresholds and can be set to generate an interrupt when the threshold is reached. This allows a 'set-and-forget' approach where you don't have to poll the sensor continually. The sub address can be set using the jumpers J81, J82 and J83.

The board accepts both the original LM75 and the LM75A in a SO-8 package.

27.1. DEVICE ADDRESS

The table below gives the base address for the LM75 / LM75A devices.

	MSB							LSB
ADDRESS	1	0	0	1	A2	A1	A0	R/W

The address has 3 user definable bits through pins on the device, allowing for a total of up to 8 sensors on the bus.

27.2. USAGE

The LM75 has an onboard 9-bit sigma delta (11-bit for the LM75A) that produces a signed number representing the measured temperature. The device can be put into a low power sleep mode by setting a bit. During sleep mode conversion is suspended but the device can be woken up be clearing the bit. Conversions are started automatically at an interval of 100mS.

The LM75 has 5 internal registers. The first register acts as a pointer into the other 4. The pointer register must be written to for every transaction.

27.2.1. Pointer register:

The pointer register allows selecting the internal register you want to operate on.

```
0000_0000 : Temperature register
0000_0001 : configuration register
0000_0010 : Hysteresis register
0000_0011 : Trip point (Tos)
```

27.2.2. Configuration Register (01)

This register allows the configuration of the device. A number of options are available.

```
Bit 0 : 0 = Active , 1 = sleep
Bit 1 : 0 = Tos acts as comparator, 1 = Tos acts as interrupt
Bit 2 : 0 = Tos is active Low , 1 = Tos is active high
Bits 3 and 4 : Tos queue programming.
```

Bit 0 determines if the device is active or in a low power sleep mode. Bit 1 selects the operation of the over temperature shutdown, sometimes called the INT or Tos pin. Bit 2 sets the polarity of this Tos pin. When the fault condition occurs the selected level will be then applied to the pin.

Bits 3 and 4 allow the programming of a digital filter. Depending upon the setting, a number of consecutive faults must occur before the output will react. This allows the filtering of spurious trips due to noise.

Setting	Queue value
00	1
01	2
10	4
11	6

27.2.3. Temperature register (00)

This register needs to be read as 2 bytes. The register itself is 16-bits long and the MSB is transmitted first. The 11-bit signed digitized temperature is stored in alignment with the MSB.

The table below shows the conversion results for the LM75A device. In the case of the original LM75 with a 9-bit converter, the precision will decrease. The last 2-bits of the 11-bit number will be zeroes.

Register contents	11-bit number	Hex	Decimal	temperature
0111 1111 0000 0000	011 1111 1000	7F3F8	1016	+127.000 °C
0111 1110 1110 0000	011 1111 0111	3F7	1015	+126.875 °C
0001 1001 0000 0000	000 1100 1000	0C8	200	+25.000 °C
0000 0000 0010 0000	000 0000 0001	001	1	+0.125 °C
0000 0000 0000 0000	000 0000 0000	000	0	0.000 °C
1111 1111 1110 0000	111 1111 1111	7FF	−1	−0.125 °C
1110 0111 0000 0000	111 0011 1000	738	−200	−25.000 °C
1100 1001 0010 0000	110 0100 1001	649	−439	−54.875 °C
1100 1001 0000 0000	110 0100 1000	648	−440	−55.000 °C

27.2.4. Trip point register (Tos) (11)

The Over temperature Shutdown (Tos) or trip point register is used to set the threshold at which the LM75 will activate its INT (Tos) output pin. This register works in conjunction with the hysteresis register. The operation of the INT pin is determined by the setting in the configuration register.

The layout of this register is identical to that of the temperature register. Two consecutive bytes must be written with the desired 9-bit value aligned to the MSB. Even though the LM75A has an 11-bit temperature range, only the top 9-bits are actually used for comparison.

27.2.5. Hysteresis register (10)

The hysteresis register allow the user to set a window (hysteresis) for the over temperature detection.

27.3. ASSEMBLY DRAWING

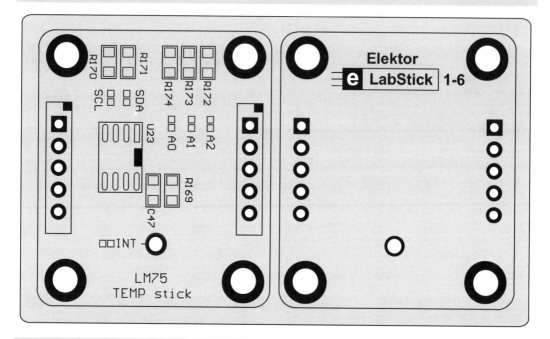

27.4. BILL OF MATERIALS

R169, R170, R171, R17, R174	10 K Ohm SMD 0805
C47	100nF 25V X7R ceramic capacitor SMD 0805
U23	LM75 in SO-8 package
K56, K57	100 mils (2.54mm) pins. Strip of 5 pins.

28. LABSTICK 1-7: PCF8563 REALTIME CLOCK

The real time clock stick allows you to add a real time clock chip to your system. The board has its own emergency power supply that preserves the operation of the clock and memory. This battery backup is achieved using a supercap of 0.47 farad (C40).

Depending upon the RTC chip installed, you can get a clock signal out or the RTC chip may provide an interrupt when a preset alarm time has passed. These signals are available on the CLK OUT and INT pins. A solder bridge allows you to feed the INT signal onto the stick connectors for linking with a CPU. Timing can be adjusted using a trimming capacitor (C42).

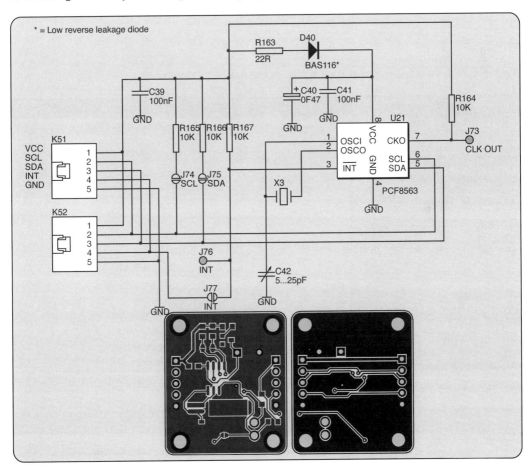

As with any I2C stick board the SCL and SDA pull up resistors can be enabled by closing the appropriate solder bridge, or they can be left open if this is not the last device in the chain.

A low leakage diode D40 allows the super capacitor to charge whenever power is applied. A series resistor R163 limits the inrush current in the case of the capacitor being empty. The power consumption of the RTC chips is so low that they can actually run for months on the energy stored in C40. The crystal is a common 32.768 KHz watch crystal in a round body. A pad is available on the printed circuit board to allow the case of the crystal to be soldered down to prevent it from possible mechanical damage.

28.1. DEVICE ADDRESS

The table below gives the base address for the PCA8563 device.

	MSB							LSB
ADDRESS	1	0	1	0	0	0	1	R/W

Note that this device has no sub addresses and only one address can be presented on an I2C bus.

28.2. USAGE

The PCF8563 has a number of registers that allow the device to be accurately controlled.

ADDRESS	ADDRESS (HEX)	FUNCTION
0	0	Control 1
1	1	Control 2
2	2	Seconds
3	3	Minutes
4	4	Hours
5	5	Days
6	6	Weekdays
7	7	Months
8	8	Years
9	9	Alarm Minutes
10	A	Alarm Hour
11	B	Alarm Day
12	C	Alarm Weekday
13	D	CLKout control
14	E	Timer Control
15	F	Timer

28.2.1. Control 1 register

Register	D7	D6	D5	D4	D3	D2	D1	D0
Control 1	TEST1	0	STOP	0	TESTC	0	0	0

The TEST1 bit allows you to run the device from an external clock. When set to 1 an external clock needs to be applied. The STOP bit halts the clock if set to 1. The CLKout keeps on running. The TESTC allows a power on reset. If set to 1 the device will not self clear at power up.

28.2.2. Control 2 register

Register	D7	D6	D5	D4	D3	D2	D1	D0
Control 1	0	0	0	TI-TP	AF	TF	AIE	TIE

TI-TP: Select INT control. If set to zero the INT pin is under control of TIE. If set to 1 the INT pin brings out a signal set by the countdown timer interrupt.

AF: alarm flag. Writing a zero clears the alarm flag. Reading it returns the alarm state.

TF: Timer flag. You can write a 0 to this location to clear the timer flag. Reading returns the current state.

AIE: alarm interrupt enabled. If set to 1 this enables the assertion of the INT pin if an alarm occurs.

TIE: timer interrupt enable. If set to 1 this allows the assertion of the It pin when a timer event occurs.

28.2.3. Seconds register

Register	D7	D6	D5	D4	D3	D2	D1	D0
Seconds	VL	T2	T1	T0	U3	U2	U1	U0

The seconds register stores the seconds counter as a 2 digit BCD encoded number. Bits U3 to U0 represent the units and bits T2 to T0 represent the tens. The data is stored as BCD which allows for easy decoding. You can simply split the byte into two nibbles, mask off the VL bit and convert to a number.

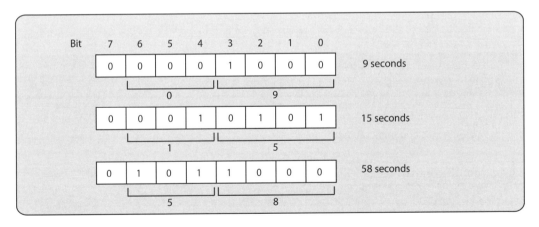

The VL flag indicates that the supply voltage has dipped below the threshold since the last read operation if it is set. If this kind of event occurs the integrity of the time and date is no longer guaranteed, and the clock should be reset.

28.2.4. Minutes register

Register	D7	D6	D5	D4	D3	D2	D1	D0
Minutes	0	T2	T1	T0	U3	U2	U1	U0

The minutes register stores the current minute counter as a two digit BCD encoded number. Decoding works on the identical principle as in the seconds counter.

28.2.5. Hours register

Register	D7	D6	D5	D4	D3	D2	D1	D0
Hours	0	0	T1	T0	U3	U2	U1	U0

The hours register stores the hour as a two digit BCD encoded number. The format is a 24 hours clock that runs from 00 to 23.-bits T1 and T0 represent the tens and bits U3 to U0 represent the units.

28.2.6. DAYS register

Register	D7	D6	D5	D4	D3	D2	D1	D0
Days	0	0	T1	T0	U3	U2	U1	U0

The days register stores the day of the month as 2 digit BCD encoded number running from 00 to 31. The PCF8563 is leap year aware. If the year counter contains a value that is exactly divisible by 4 it will add a 29th day in February. It will also do this for the year 00. Since centuries are not counted it cannot compensate for the 400 year exception rule.

28.2.7. Weekdays register

Register	D7	D6	D5	D4	D3	D2	D1	D0
Weekdays	0	0	0	0	0	U2	U1	U0

This register stores the day of the week as a single BCD number ranging from 0 to 6. The weekday begins on Sunday (0) and ends on Saturday (6).

28.2.8. Months register

Register	D7	D6	D5	D4	D3	D2	D1	D0
Months	C	0	0	T0	U3	U2	U1	U0

The months register stores the month as a digit BCD encoded number ranging from 01 to 12. The C bit is a century flag. If this flag is set it means we are in the 21st century (years 20xx). If cleared the clock operates in the 20th century (19xx).

28.2.9. Years register

Register	D7	D6	D5	D4	D3	D2	D1	D0
Years	T3	T2	T1	T0	U3	U2	U1	U0

The years register stores the tens and units of the year as a two digit BCD number ranging from 00 to 99.

28.2.10. Alarm Minutes register

Register	D7	D6	D5	D4	D3	D2	D1	D0
A Minutes	AE	T2	T1	T0	U3	U2	U1	U0

This register sets the minutes 'compare' value for the alarm function. The contents of the tens and units bits are compared with those of the minutes register. If the AE bit in this register is set this register is taken into account.

28.2.11. Alarm Hour register

Register	D7	D6	D5	D4	D3	D2	D1	D0
A hours	AE	0	T1	T0	U3	U2	U1	U0

This register sets the hours 'compare' value for the alarm function. This register can be enabled by setting the AE flag.

28.2.12. Alarm DAY register

Register	D7	D6	D5	D4	D3	D2	D1	D0
A Day	AE	0	T1	T0	U3	U2	U1	U0

This register sets the alarm days 'compare' value for the alarm function. This register can be enabled by setting the AE flag.

28.2.13. Alarm Weekday register

Register	D7	D6	D5	D4	D3	D2	D1	D0
A weekday	AE	0	0	0	0	U2	U1	U0

This register sets the alarm weekdays 'compare' value for the alarm function. This register can be enabled by setting the AE flag.

28.2.14. CLK out control register

Register	D7	D6	D5	D4	D3	D2	D1	D0
CLK out	FE	0	0	0	0	0	D1	D0

This register controls the CLK OUT signal of the PCA8563. If the FE bit is set to 1 the CLK Out signal is active. If the bit is cleared, the pin is switched into a high impedance mode.

Setting	CLK-OUT frequency
00	32.768 KHz
10	1024 Hz
01	32 Hz
11	1 Hz

Depending on the settings of D1 and D0 the frequency available at CLKout can be one of four possible rates.

28.2.15. Timer control register

Register	D7	D6	D5	D4	D3	D2	D1	D0
A Minutes	TE	0	0	0	0	0	TD1	TD0

The timer control register allows operation of the timer system in the PCA8563. The TE bit enables the timer to run if it is set to 1. Depending upon the settings in the control register 2, the timer will assert the INT pin when it reaches zero. If the timer is preloaded with a value, it will count down until it hits zero. The TD1 and TD0 bits allow the selection of a clock source for the timer.

Setting	CLK-OUT frequency
00	4096Hz
10	64Hz
01	1 Hz
11	1/60 Hz (1 tick per minute)

28.2.16. Timer register

Register	D7	D6	D5	D4	D3	D2	D1	D0
A Minutes	T7	T6	T5	T4	T3	T2	T1	T0

The timer register can be used for various purposes, it offers a pre-settable count-down—, that can generate an interrupt. This timer can be used as a watchdog or as a sleep timer.

28.3. ALARM OPERATION

The PCA8563 has an elaborate alarm mechanism that deserves some additional attention. Each individual parameter (minutes, hours, days and weekdays) can be pulled in to the alarm system.

By setting the alarm enable bit (AE) in an alarm register the parameter becomes active in the alarm comparator.

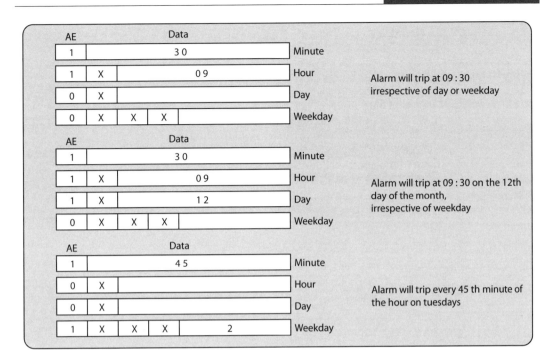

28.4. ASSEMBLY DRAWING

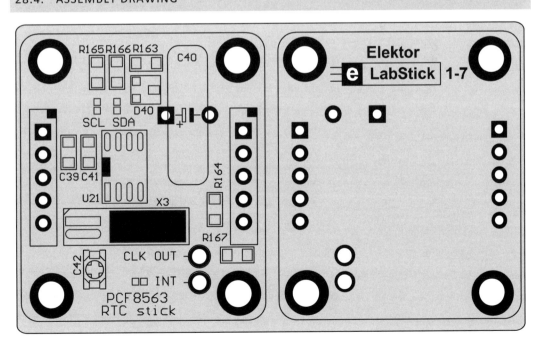

28.5. BILL OF MATERIALS

R165, R166, R164, R167	10K ohm SMD 0805
R163	22 ohm SMD 0805
C40	0.47F 5.5V vertical supercap
C39, C41	100nF 25V X7R ceramic capacitor SMD 0805
C42	22pF trim capacitor SMD
D40	BAT54C Schottky diode SOT23
U21	PCF8563 / PCF8583 RTC
K51, K52	100 mils (2.54mm) pins. Strip of 5 pins.
X3	32.768 KHz watch crystal

29. LABSTICK 1-8: 8-BIT PROTECTED OUTPUT

Besides protected input circuitry it may also be beneficial to have a kind of robust output that can handle more current than the average I/O expander and that is exactly what this module provides. This module has a counterpart that provides an 8-bit protected input.

A screw terminal connector provides easy access to the outgoing signals as well as ground and the fly back diode's common point. Each output terminal has a corresponding LED that shows the ON or OFF state. Keep in mind that in this case an ON output means the pin is connected to GROUND.

For this implementation, you must use the PCA9534 or similar, as this circuitry needs the push pull capabilities of those chips in order to correctly drive the LED's and the ULN2803. This circuitry will not work with a PCA8574 or any other non push pull I/O driver. Moreover, the PCF8574 will power up with its outputs set high which may have an unintended effect when all of the ULN2803 outputs switch on simultaneously.

The entire schematic is pretty much a 'cookie cutter' application as the usual address selection mechanism and selectable pull up resistors for SDA and SCL have both been implemented.

A ULN2803 provides the output signals. This device can sink up to 500mA per output pin and has snubber diodes that protect the output against back EMF when driving inductive loads such as relays.

Even though each output can handle 500mA the total current of the device is limited to 2.5 ampere. This is something that is commonly overlooked. The common terminal of the ULN2803 cannot have more than 2.5 ampere simply because the internal bonding wires cannot handle it and the internal dissipation becomes too high.

If you want to use all 8 outputs simultaneously you should not use more than 300mA per output.

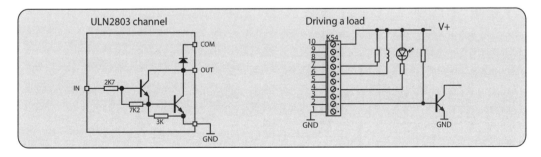

LabWorX LabStick 1-8: 8-Bit Protected Output

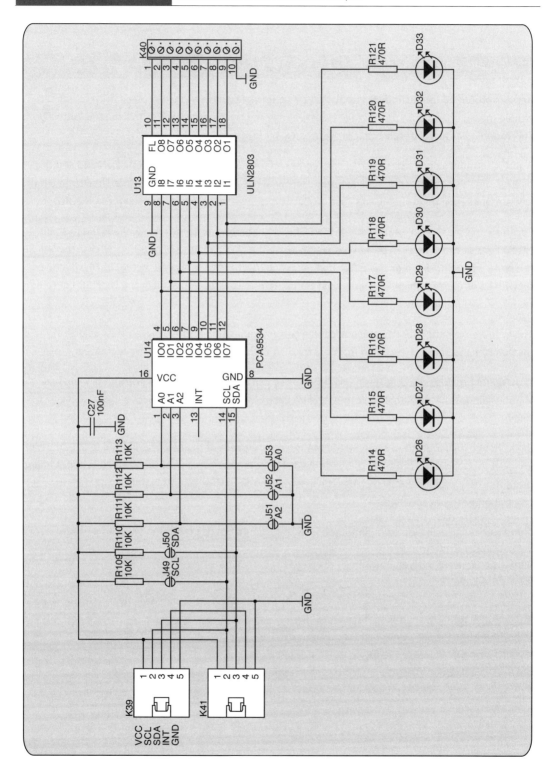

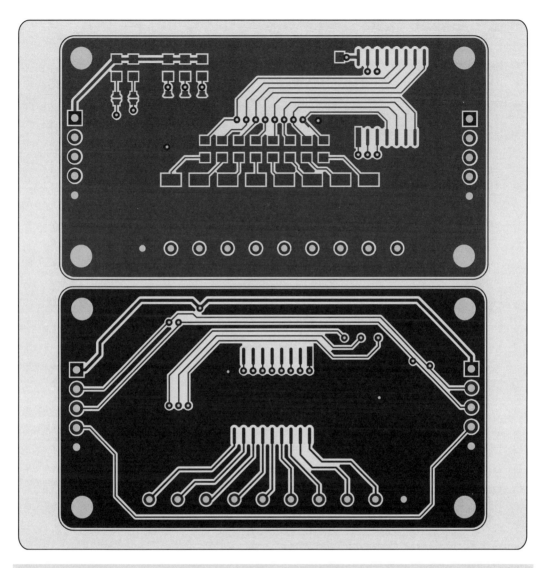

29.1. ASSEMBLY

Start by installing all the surface mount components on the top. Next install the connectors on the front before flipping the board over and soldering them in. Once the connectors are installed, you can then install the ULN2804 on the back.

LabWorX LabStick 1-8: 8-Bit Protected Output

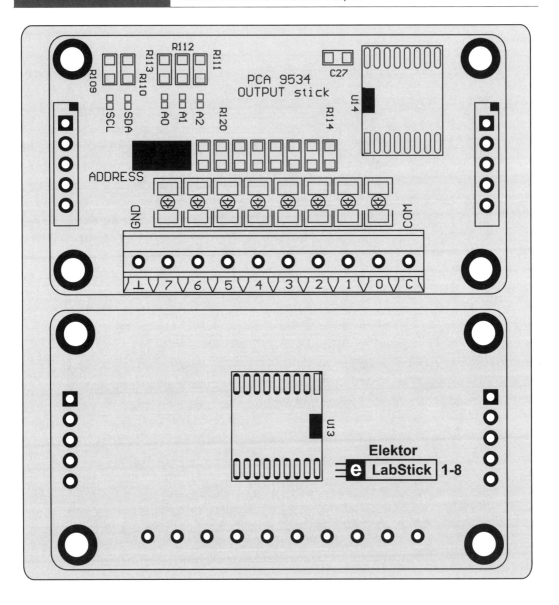

29.2. BILL OF MATERIALS

R109, R110, R111, R112, R113	10 K Ohm SMD 0805
R114, R115, R116, R117, R118, R119, R120, R121	22 Ohm SMD 0805
C27	100nF 25V X7R ceramic capacitor SMD 0805
D26, D27, D28, D29, D30, D31, D32, D33	LED, Colour of choice PLCC-2 PACKAGE (TOPLED)
U14	PCA9534
U13	ULN2803
K39, K41	100 mils (2.54mm) pins. Strip of 5 pins.
K40	TERMINAL BLOCK

30. LABSTICK 1-9: 8-BIT PROTECTED INPUT

So far I have only dealt with basic I2C systems that can live in a shielded environment. Once we start dealing with the 'off board' world things tend to get rough very quickly. All these integrated circuits are happy in their 5 or 3.3 volt world with nice, clean, signals. And then someone goes and upsets this by connecting bouncy mechanical switches, noisy signals, large voltages and even induced currents because of cabling.

So we need to come up with some interface circuitry to condition the signals and protect our electronics from the outside world.

This protected input board is such a module. It has a counterpart output module that was covered earlier in this book.

30.1. SCHEMATIC

The schematic is given on the next page and merits some explanation as not everything will be clear.

Let's start at the connector K54. This is a screw terminal type connector that allows for easy connection to 'industrial grade' hardware such as pushbuttons, contactors, relays and the likes. The resistors R91...94 and R97 to R100 in combination with jumpers J47 and J48 allow you to set the inputs to pull up or down.

In the case of pull down, the pins of the I/O chip will be read as logic 0 and you must apply a voltage larger than 1.5 volts to the input to set it high. The connector K37 has outgoing power available through diode D13 for that purpose. If you close J48 the default mode is pull up. In this case an open pin will be read as logic HIGH and you must pull the input to ground externally to read a 0. Ground is available on K37 as well.

The diode D13 protects against voltages above VCC that may be applied to the connector. It is perfectly fine to inject an external voltage (for example 12 volts) and use that as pull up voltage for the resistors R91...94 and R97 to R100.

The signals coming from the input now go through resistors R72, R74, R82, R84, R86, R88, R90 or R96 and pass the double diodes D8, D10, D11, D12, D14, D15, D16 or D17. Any signal lower than ground – 0.7 volts will be clamped by the bottom half of the diode. Any signal above (Veneer of D9) + 0.7 will be clamped at that level. Once the signal has been restricted in positive and negative amplitude it passes through R71, R73, R81, R83, R85, R87, R89 or R95 and enters the I2C I/O expander.

LabStick 1-9: 8-bit Protected Input

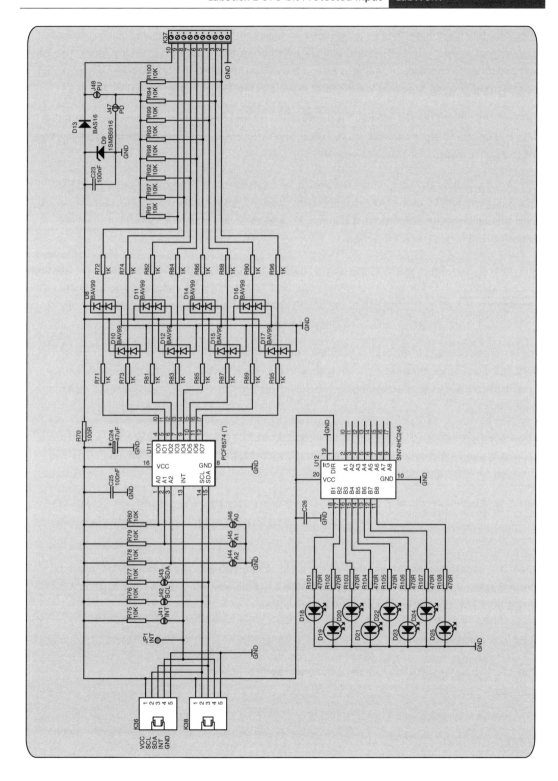

I2C Bus

A logic buffer chip is connected to the cleaned up signals and drives LED's that show the state of each individual input. The LED's have been placed adjacent to the correct pin on the connector. This aids in debugging the system cabling as well as providing the necessary 'blinking-lights' that attract people to the system. Note that the PCF8574 may detect logic high earlier than the 74HC245.

The circuitry around the I/O expander is pretty much cut and dried. Three jumpers in combination with pull up resistors provide address selection, and three other jumpers provide the capability to inject a pull up resistor on SDA, SCL and INT.

The circuitry around the zener diode requires some explanation. If an incoming signal has a level that is larger than VCC of the I/O expander there is risk for potential damage to the input structure. Even though integrated circuits have clamping diodes, these are made for ESD reasons and not really for continuous current injection.

The zener diode D9 is selected in such a way that it creates a reference voltage which is 1 diode drop (0.7 volts) below VCC. For a 5 volt system this zener diode is set to 4.3 volt, for a 3.3 volt system this zener can be 2.8 volts. The top half of the double diodes BAV99 will go into conduction for any voltage that is higher than Vzener + Vf of the BAV99.

For fast transients C23 forms a low impedance path to ground. This aids in further dampening sharp voltage spikes. Zener diodes are relatively slow to turn on. R70 pre biases the zener diode with a small current. This keeps the zener diode already in an 'on' state so it will react faster to incoming over voltages.

The protection is twofold. The large jolt is current limited by the incoming resistors, clamped and clipped by the BAV99 double diodes to a safe voltage provided by the zener diode. An AC path is provided for fast transients.

Any residual energy remaining at this point is further reduced by the series resistors between the diode clamp and the IC input. This rest energy is then dispersed in the internal protection diodes of the I/O expander.

The I/O expander chip can be any device that is pin compatible with the PCF8574. Care must be taken that the internal weak pull ups do not cause any problems when pulling the signal adequately low. A device without internal pull ups will work better.

30.2. CIRCUIT BOARD

The next two pages show the copper layout and assembly plan of the board.

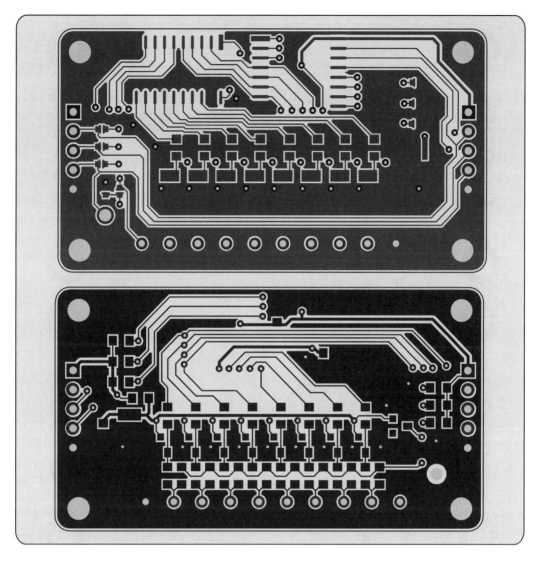

When assembling, start with the surface mounted parts on the back first but leave the zener diode for later. Next, mount the components on the front including the capacitor. Install the connectors, flip the board over and solder the connectors in place. Now you can install the zener diode. The thickness of the zener diode would have made the board wobbly when working on the front, if you had soldered it in before.

Take care when ordering the 74245 that you get the 'skinny' version. These devices are available in two widths and this board needs the narrow one.

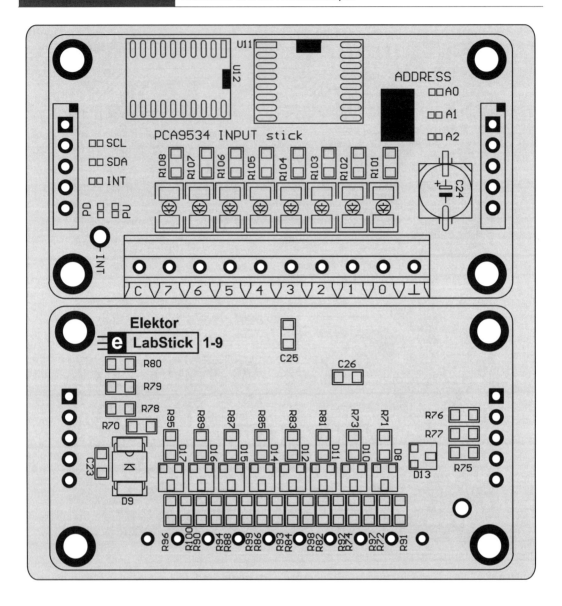

30.3. BILL OF MATERIALS

R71, R72, R73, R74, R81, R82, R93, R84, R85, R86, R87, R88, R89, R90, R95, R96	1K Ohm SMD 0805
R75, R76, R77, R78, R79, R80, R91, R92, R93, R94, R97, R98, R99, R100	10K Ohm SMD 0805
R101, R102, R103, R104, R105, R106, R107, R108	470 Ohm SMD 0805
R70	100 Ohm SMD 0805
C23, C25, C26	100nF 25V X7R ceramic capacitor SMD 0805
C24	47uF 16V Electrolytic
D9	Zener diode 4.3V SMB package
D8, D10, D11, D12, D14, D15, D16, D17	BAV99 OR BAT54C DUAL DIODE SOT-23
D13	MMBD4148 in SOT23
D18, D19, D20, D21, D22, D23, D24, D25	LED PLCC 2 (TOPLED) PACKAGE. Colour to your liking (Red, Green, Orange or Yellow. White and Blue can be used if the board runs on 5 volts.
U11	PCF8574 or pin compatible I/O expander
U12	74HC245 or 74HCT245 in SO20 narrow package (5.3mm) Made by Texas Instruments, ON semi and Fairchild.
K36, K38	100 mils (2.54mm) pins. Strip of 5 pins.
K54	TERMINAL BLOCK

30.4. SOME APPLICATION INFORMATION

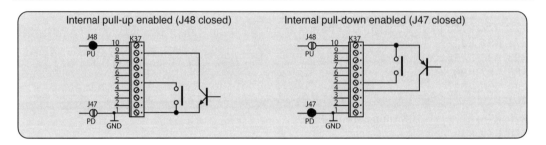

The left side shows the condition when active low signals will be used to drive the input circuitry. The right image shows the same but this time for active high signals.

31. LABSTICK 1-10: MCP4725 D/A CONVERTERS

The MCP4725Ax are small 12-bit D/A converters that allow placement directly at the point where the signal is needed. The devices have an onboard EEprom that can retain the settings of the DAC. After a power cycle the DAC value is restored at the output. These devices are ideal for digital trimming applications in a system.

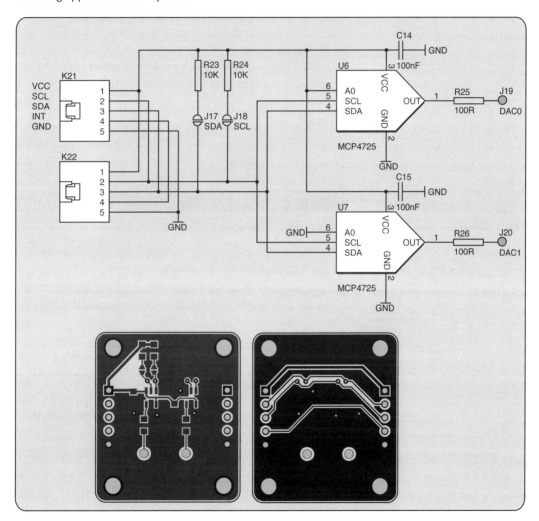

The application board has the classic interface connectors and the selectable pull up resistors on SDA and SCL. These devices come in a SOT23-6 package. Since these are so small there was room on the stick to provide two of them. The stick pre-wires the devices to one of two possible slave addresses.

The MCP4725 devices can work in standard (100 KHz), Fast (400 KHz) and high speed (3.4 MHz) mode and are compatible with both 3.3 and 5 volt operations and can work all the way down to 2.7 volts.

There is no internal reference. Instead, the power supply is used as the reference for the DAC. The output value is thus a fraction of the applied supply voltage.

Vout = (Vcc / 4096) * data

Since VCC is used as a reference this supply should be clean. Any noise present will be fractionally divided at the output.

The onboard EEprom can store the current DAC settings and recall them after a power cycle. The device will also respond to the I2C General call – Reset command, upon which it will also reapply the stored value to the DAC.

31.1. DEVICE ADDRESS

The table below gives the base address for the MCP4725 devices.

	MSB							LSB
ADDRESS	1	1	0	0	A2	A1	A0	R/W

The MCP4725 has three bits that determine the slave address of the devices. Bits A2 and A1 are programmed in the factory, depending upon the order code of the part.

Part Number	Address (A2 and A1)	Device marking
MCP4725A0T	00	AJNN
MCP4725A1T	01	APNN
MCP4725A2T	10	AQNN
MCP4725A3T	11	ARNN

Bit A0 of the device address is available as a pin on the device and can be selected by the user. The combination of the factory programmed and user address bits allows for a total of 8 devices on the same bus.

31.2. USAGE

The data layout in normal or in high speed mode is different. I will focus on the normal mode of operation only since most masters are not capable of running in high speed mode. The high speed mode is explained in detail in the Microchip datasheet.

The device requires 3 bytes to be written after the addressing for each transfer.

31.2.1. Control byte

The first byte is the control register.

	MSB							LSB
Control Byte	C2	C1	C0	X	X	PD1	PD0	X

The three command bits (C2, C1 and C0) specify the operation that you want to perform. There are only two possible settings.

- 010: writes to the DAC but does NOT store the setting in the EEPROM. This allows for fast manipulation of the DAC without wearing out the EEPROM device. Once a correct setting has been found, or you want to store the power up value you can issue the second command.
- 011: writes to both the DAC and EEPROM. The settings will be retained and restored upon a power up or general call for reset on the I2C bus.

The power down control bits (PD1 and PD0) select the mode of operation for the device:

- 00: normal operation. The DAC output is active.
- 01: sleep mode. The DAC is suspended and the output is connected via 1K to ground.
- 10: sleep mode. The DAC is suspended and the output is connected with 100K to ground.
- 11: sleep mode. The DAC is suspended and the output is connected via 500K to ground.

Unused bits are marked with an x and are 'don't care'. It is not advisable to send other command codes than 010 and 011 in the command bits as these are reserved for 'future use', therefore the function is unknown at this point.

31.2.2. Data bytes

The control byte is followed by two more data bytes containing the actual data to be put in the DAC. The DAC value is a 12-bit unsigned number aligned with the MSB.

Bit	15	14	13	12	11	10	9	8	7	6	5	4	3	2	1	0
Data	D11	D10	D9	D8	D7	D6	D5	D4	D3	D2	D1	D0	X	X	X	X

31.2.3. Write datagram

The diagram below shows the contents of a full datagram for a typical write operation.

Byte	Direction	D7	D6	D5	D4	D3	D2	D1	D0	CONTENT
0	OUT	1	1	0	0	A2	A1	A0	0	ADDRESS
1	OUT	C2	C1	C0	X	X	PD1	PD0	X	STATUS
2	OUT	D11	D10	D9	D8	D7	D6	D5	D4	DAC DATA
3	OUT	D3	D2	D1	D0	X	X	X	X	DAC DATA

31.3. READ OPERATIONS

The device is not symmetrical between read and write operation. A read operation will return a status byte first

	MSB							LSB
Control Byte	RDY	POR	X	X	X	PD1	PD0	X

The RDY bit shows that the EEPROM write operation has completed when set to 1. If this bit is 0 then EEPROM writing is still active.

The POR bit shows that a power on reset event has occurred and that the DAC value and operation has been reset to the stored settings.

PD1 and PD0 return the current settings for the power down control bits.

Subsequent bytes return the current settings of the DAC (2 bytes containing a 12-bit number aligned with the MSB of the 2 bytes.) and 2 more bytes return the contents of the EEPROM memory.

31.3.1. Read datagram

Byte	Direction	D7	D6	D5	D4	D3	D2	D1	D0	CONTENT
0	OUT	1	1	0	0	A2	A1	A0	1	ADDRESS
1	IN	RDY	POR	X	X	X	PD1	PD0	X	STATUS
2	IN	D11	D10	D9	D8	D7	D6	D5	D4	DAC DATA
3	IN	D3	D2	D1	D0	X	X	X	X	DAC DATA
4	IN	X	PD1	PD0	X	D11	D10	D9	D8	EEPROM DATA
5	IN	D7	D6	D5	D4	D3	D2	D1	D0	EEPROM DATA

31.4. ASSEMBLY DRAWING

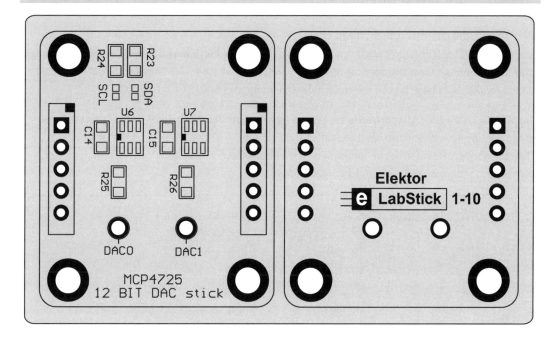

31.5. BILL OF MATERIALS

R23, R24	10 K Ohm SMD 0805
R25, R26	100 Ohm SMD 0805
C14, C15	100nF 25V X7R ceramic capacitor SMD 0805
U6, U7	MCP4725 SOT23-6 PACKAGE
K21, K22	100 mils (2.54mm) pins. Strip of 5 pins.

32. LABSTICK 1-11: ADC081 / ADC101 / ADC121 A/D CONVERTERS

Even though the PCF8591 is also an A/D converter it is interesting to have a different device in the arsenal. The PCF8591 is an older device and comes in a rather bulky package and the 8-bit resolution may not be enough for some applications. There are plenty of other I2C compatible A/D converters on the market but I have picked the National semiconductors ADC121 for this stick.

The ADC121 is a 12-bit A/D converter that is available in both MSOP and SOT23-6 packages. These are really tiny devices and are ideal to scatter throughout the design, directly at the point where they are needed. No need to run long analog traces all through the system, just drop in the converter on the spot and tie it on to the I2C network.

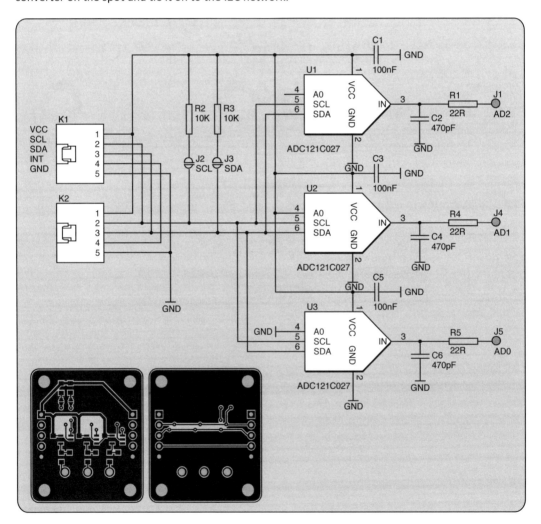

If less resolution is acceptable the ADC081 device can be used as well. These are pin and function compatible but offer only 8-bit resolution. There is also a 10-bit pin compatible version available under the name ADC101.

For this board I have used the ADC121 in SOT23-6 package. This 6 pin device has only power and ground, an input, the mandatory I2C lines SCL and SDA and one address selection pin. If more address pins are required then the devices in MSOP 8 package have these available.

Since these devices are so small, the stick allows you to install 3 of them. The address selection pin is not a standard digital pin. There are 3 possible sub addresses for this device and the selection is made by either tying the A0 to power, ground or leaving it open.

Each A/D converter has its own little input filter that blocks high frequency noise.

The ADC121 family devices can work in standard (100 KHz), fast (400 KHz) and high speed (3.4 MHz) mode and are compatible with both 3.3 and 5 volt operations and can work all the way down to 2.7 volts.

One variation has an ALERT output instead of the regular address pin.

resolution	SO23 with ALERT pin	SOT23 with ADDR pin
12-bit	ADC121C021	ADC121C027
10-bit	ADC101C021	ADC101C027
8-bit	ADC081C021	ADC081C027

The board supports only the C027 variation that features the address pin. Do not install the C021 version as addressing will not work and the devices may be damaged due to a hardwired alert output.

32.1. DEVICE ADDRESS

The table below gives the base address for the ADCxxxC027 devices.

	MSB							LSB
ADDRESS	1	0	1	0	0	A1	A0	R/W

Only the combinations 00, 01 and 10 are valid for A1 and A0. The combination 11 is reserved for the ADCxxxC021.

The devices in MSOP8 have access to more possible addresses. Consult the datasheet for more information on those devices.

32.2. USAGE

The ADCxxx devices have a number of built in registers that allow for several functions. The registers are accessible through a selector register.

32.2.1. Selector register

Bit	7 (MSB)	6	5	4	3	2	1	0 (LSB)
ADDRESS	0	0	0	0	0	R2	R1	R0

The bits R2 to R0 select one of a possible 8 registers in the device

Register	R2	R1	R0
Conversion result (read only)	0	0	0
Alert status (read/write)	0	0	1
Configuration (read/write)	0	1	0
Low Limit (read/write)	0	1	1
High limit (read/write)	1	0	0
Hysteresis (Read/Write)	1	0	1
Lowest conversion (Read/Write)	1	1	0
Highest Conversion (Read/Write)	1	1	1

32.2.2. Conversion result register

The result of the A/D conversion is available through this register. Upon reading this register a new conversion will automatically start if the device is in normal operation mode. The data is stored as a 16 bit number and thus 2 bytes need to be read from the I2C bus to obtain the complete result.

Bit	15	14	13	12	11	10	9	8	7	6	5	4	3	2	1	0
Data	ALERT	0	0	0	D11	D10	D9	D8	D7	D6	D5	D4	D3	D2	D1	D0

The alert bit signifies that an alert condition has occurred. Alert conditions are set up in the configuration register.

Bits 14 through 12 are reserved and will return logic zero. Bits 11 through 0 contain the result of the conversion. Of course, if you use the ADC081 or ADC101 this number of bits will be 8 or 10. Leading bits will be padded with zeroes. In other words: the data is always aligned on the LSB.

32.2.3. Alert status register

This register holds information on the source of the alert. If the ALERT bit is set in the Conversion result register, then the alert register can be read to identify the nature of the event.

Bit	7 (MSB)	6	5	4	3	2	1	0 (LSB)
ADDRESS	0	0	0	0	0	0	OVER	UNDER

The OVER range bit is set when the input value has exceeded the limit set in the High limit register. The UNDER range bit will be set if the input has fallen below the value set in the under range register.

To clear the bits simply write 00 in this register.

32.2.4. Configuration register

The configuration register sets the options for the converter system

Bit	7 (MSB)	6	5	4	3	2	1	0 (LSB)
ADDRESS	C2	C1	C0	ALERT HOLD	AFE	APE	0	POLARITY

- Bits C2 to C0 allow the setting of the A/D measurement interval. If set to 000 the A/D converter is put into sleep mode. Conversion will only be initiated after a read from the ADC conversion register. Keep in mind that you always get the result of the previous conversion. In order to get 'fresh' data you need to perform two consecutive reads. The first read will trigger the conversion to start and the second will retrieve the result. Make sure you leave enough time between the read operations to allow the converter to finish a conversion. To eliminate noise it is advisable not to perform any operation on the I2C bus during the actual conversion time. Other combinations for C2 to C0 set an interval based on the internal clock of the ADCxxx.
- The Alert Hold bit controls the behaviour of the alert bits in the alert status register. If the Alert Hold bit is cleared (0), then the alert bits (over and under) will self clear when the voltage comes on again within the preset range. If the Alert hold bit is set (1) the bits will not clear. This mode allows you to detect that an over- or under-range has occurred between the last time you cleared the Alert Status register and now.
- APE or Alert Pin enable is of no consequence for the devices on our board. The devices that have an Alert output pin do use this. If the APE bit is set to 0 the ALERT output is disabled and be in a high impedance mode. If the bit is set then the ALERT flag will trip the ALERT output.
- AFE Alert Flag enable. This bit enables or disables the ALERT bit in the status register. If the AFE bit is set to zero no alerts will be signalled in the ALERT bit. If the AFE bit is set then the tripping of either UNDER or OVER range will also set the ALERT bit.
- Polarity allows you to specify if the ALERT output is to be active high (bit is 1) or active low (bit is 0). For the devices on our board this is of no consequence.

Sampling speed (bits C2 to c0)

Speed	C2	C1	C0
Disabled	0	0	0
27 Ks/s	0	0	1
13.5 Ks/s	0	1	0
6.7 Ks/s	0	1	1

Speed	C2	C1	C0
3.4 Ks/s	1	0	0
1.7 Ks/s	1	0	1
0.9 Ks/s	1	1	0
0.4 Ks/s	1	1	1

32.2.5. Low limit (underrange) trip point

This register sets the lower limit for the alert detection. Whenever the digitized voltage falls below the value set in this register the UNDER alert flag will trip. The data is stored as a 16-bit number so two consecutive bytes need to be written to completely store the value. The data is aligned to the LSB. The four MSB's must be set to all 0.

Bit	15	14	13	12	11	10	9	8	7	6	5	4	3	2	1	0
Data	0	0	0	0	D11	D10	D9	D8	D7	D6	D5	D4	D3	D2	D1	D0

32.2.6. High limit (overrange) trip point

This register sets the upper limit for the alert detection. Whenever the digitized voltage rises above the value set in this register the OVER alert flag will trip. The data is stored as a 16-bit number so, as with the lower limit register, two consecutive bytes need to be written to completely store the value. The data is aligned to the LSB. The four MSB's must be set to all 0.

Bit	15	14	13	12	11	10	9	8	7	6	5	4	3	2	1	0
Data	0	0	0	0	D11	D10	D9	D8	D7	D6	D5	D4	D3	D2	D1	D0

32.2.7. Alert hysteresis register.

This allows the setting of a hysteresis around the high and low trip points. Whenever either trip point has been passed, the value must first return to a level that is outside the hysteresis band. For example, if the high trip point is set to 2 volts, the low trip point is set to 1 volt and the hysteresis is set to 0.2 volts, and assuming that the self-clearing condition has been setup in the configuration register, the following will happen. Upon passing the 2 volts trip point the HIGH alert bit will trip. Only when the input voltage has fallen below 1.8 volts (high trip point – hysteresis) will this error clear. If the voltage goes below 1 volt the UNDER flag will trip. The voltage needs to rise to 1.2 volts (low trip point + hysteresis) before the flag will clear.

This is also a 16-bit value that needs to be written as two consecutive bytes

Bit	15	14	13	12	11	10	9	8	7	6	5	4	3	2	1	0
Data	0	0	0	0	D11	D10	D9	D8	D7	D6	D5	D4	D3	D2	D1	D0

32.2.8. Highest and lowest conversion registers

These 16-bit registers hold the lowest and highest value that was detected between the last clearing of these registers and now. These registers are only active when automatic conversion is enabled (C2 to C0 bits in the configuration register). In manual mode, these registers are not updated.

The HIGH register will be erased by writing 0x0000 to it. The lowest register must be erased by writing 0x0FFF to it.

Bit	15	14	13	12	11	10	9	8	7	6	5	4	3	2	1	0
Data	0	0	0	0	D11	D10	D9	D8	D7	D6	D5	D4	D3	D2	D1	D0

As with the other 16-bit registers the value is aligned on the LSB and the leading 4 bits must remain zero.

32.3. WRITE OPERATIONS

A write operation must always include the pointer register followed by the payload.

32.3.1. 8-bit write

Byte	Direction	D7	D6	D5	D4	D3	D2	D1	D0	CONTENT
0	OUT	1	0	1	0	0	A1	A0	0	ADDRESS
1	OUT	0	0	0	0	0	P2	P1	P1	POINTER
2	OUT	D7	D6	D5	D4	D3	D2	D1	D0	DATA

32.3.2. 16-bit write

Byte	Direction	D7	D6	D5	D4	D3	D2	D1	D0	CONTENT
0	OUT	1	0	1	0	0	A1	A0	0	ADDRESS
1	OUT	0	0	0	0	0	P2	P1	P1	POINTER
2	OUT	D15	D14	D13	D12	D11	D10	D9	D8	DATA
3	OUT	D7	D6	D5	D4	D3	D2	D1	D0	DATA

32.4. READ OPERATIONS

Prior to executing a read operation, a write operation must be performed to set the register pointer. Either a standalone write can be done, followed by stop, start and the address in read mode; or a repeated start can be given.

32.4.1. Set pointer

Byte	Direction	D7	D6	D5	D4	D3	D2	D1	D0	CONTENT
0	OUT	1	0	1	0	0	A1	A0	0	ADDRESS
1	OUT	0	0	0	0	0	P2	P1	P1	POINTER

32.4.2. Retrieve 8-bit

Byte	Direction	D7	D6	D5	D4	D3	D2	D1	D0	CONTENT
0	OUT	1	0	1	0	0	A1	A0	1	ADDRESS
1	IN	D7	D6	D5	D4	D3	D2	D1	D0	DATA

32.4.3. Retrieve 16-bit

Byte	Direction	D7	D6	D5	D4	D3	D2	D1	D0	CONTENT
0	OUT	1	0	1	0	0	A1	A0	1	ADDRESS
1	IN	D15	D14	D13	D12	D11	D10	D9	D8	DATA
2	IN	D7	D6	D5	D4	D3	D2	D1	D0	DATA

32.5. ASSEMBLY DRAWING

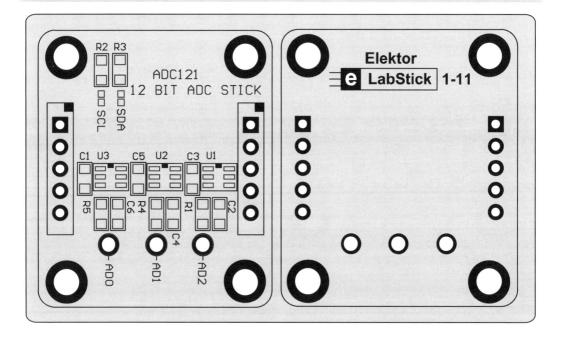

32.6. BILL OF MATERIALS

R2, R3	10 K Ohm SMD 0805
R1, R4, R5	22 Ohm SMD 0805
C1, C3, C5	100nF 25V X7R ceramic capacitor SMD 0805
C2, C4, C6	470pF 50V C0G / NP0 SMD 0805
U1, U2, U3	ADC121C027 / ADC101C027 or ADC081C027 SOT23-6 PACKAGE
K1, K2	100 mils (2.54mm) pins. Strip of 5 pins.

33. LABSTICK 1-12 PCF8591 COMBINED A/D AND D/A CONVERTER

The PCF8591 is one of the oldest I2C compatible chips and contains an 8-bit DAC and 4 analog inputs / comparators. Using the DAC and the comparators a successive approximation ADC has been constructed that the PCF8591 can digitize 4 independent channels. A control register allows you to switch the input configuration from single ended to differential. The DAC is an R-2R construction and the device gives you access to both the top and bottom of the ladder chain using the AGND and VREF pins.

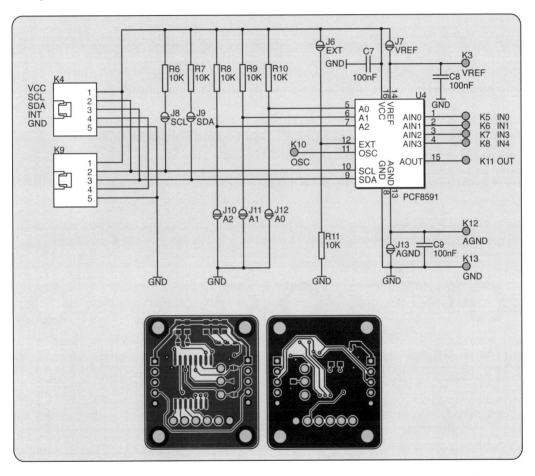

As usual, the SCL and SDA have pull ups that can be engaged by closing J8 and J9, and 3 address lines that can be set using J10, J11 and J12. The Oscillator input pin is available as test point K10, and the PCF8591 can be set to use this external oscillator by closing jumper J6.

The top and bottom of the R-2R ladder DAC are available on K3 and K12. For most applications Jumpers J7 and J13 need to be closed. This applies the power and ground and allows the DAC output to swing between ground and 255/256 of VCC. If a different range is required then J7 can be opened, and a reference voltage applied to K3. If the bottom value of the DAC needs to be lifted, then J13 can be opened and the low reference voltage applied to K12.

Suppose you want a DAC that has an output between 1 and 2 volts. You would open J7 and J13. By now applying 1 volt to K12 you set the low end of the DAC range. By applying 2 volts on K3 (VREF) you set the high end of the range. It is import to make sure that neither AGND nor VREF go below ground, or above VCC of the chip. It is not recommended to set AGND to voltages higher than those present at VREF. As the ladder is a pure R-2R this should be possible, but the chip may not like this internally.

33.1. PCA9691

The PCA9591 is a pin and function compatible CMOS superset of the older PCF8591. The key difference is in the addressing range available and the fast mode + capability of the PCA9691.

The address field is programmable from 0x40 to 0xBE to allow up to 64 devices on a single I2C bus. The datasheets have the full addressing table. To allow the extra combinations the address pins must be tied to various combinations of VCC, GND SDA and SCL.

The PCA9691 can be installed on this stick but the extra address combinations will have to be jumper wired.

33.2. ASSEMBLY DRAWING

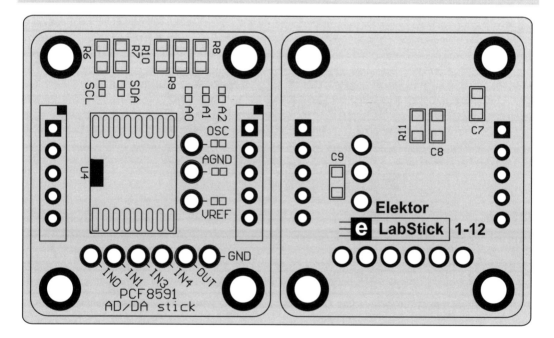

33.3. BILL OF MATERIALS

R6, R7, R8, R9, R10, R11	10 K Ohm SMD 0805
C7, C8, C9	100nF 25V X7R ceramic capacitor SMD 0805
U4	PCF8591
K4, K9	100 mils (2.54mm) pins. Strip of 5 pins.

34. LABSTICK 1-13: POTENTIOMETER

There are some applications where it is desirable to have digital control over an analog signal. Of course you can use a D/A converter but that only gives you voltage control. To set a gain you are better off with a programmable resistor that can be used as the feedback element. Of course you can make a voltage controlled amplifier, or multiplier, but those tend to be pretty complex circuits.

The MCP40D1x and MCP401x devices are available in a number of resistance values and pin-out configurations. I have opted to go with the 17 variant which has a floating resistor available.

The MCP4017, MCP4018, MCP4019, MCP40D18 and MCP40D19 can be installed on this board as well, but they do have different behaviours.

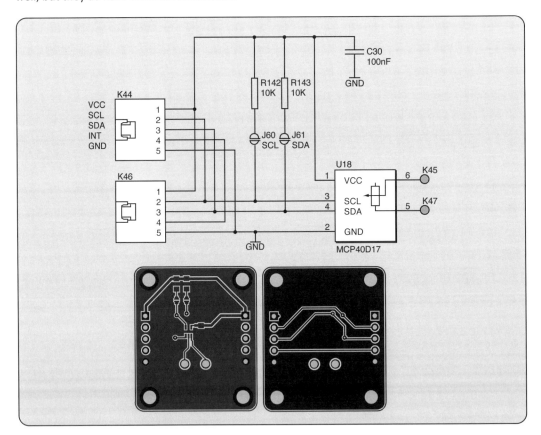

The board has the common features for the LabStick modules. Connectors K44 and K46 give the familiar interface. The jumpers J60 and J61 give the possibility to enable pull-up resistors on the bus if so desired.

There are no option jumpers to set a device address as these devices do not support such feature.

34.1. FUNCTIONAL DIFFERENCES

The MCP40xx family has 3 base configurations for the potentiometer function.

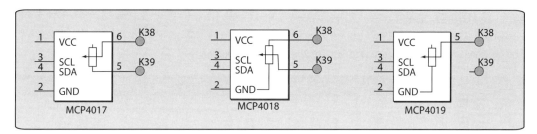

The MCP4017 offers a fully isolated adjustable resistor. Only one end and the wiper are brought out. The other resistor end is not connected. The MCP4018 places the resistor ladder between pin 6 and ground of the device. The wiper comes out just as with the MCP4017.

The MCP4019 only brings out the wiper and provides a resistive path to ground.

The MCP4017 and MCP4019 are so called rheostat configurations while the MCP4018 is a true ratiometric operation.

The devices power up in mid scale and do not retain their settings. They are available in a range of resistive values.

While these devices behave like an adjustable resistor they are made from a switched resistor network. The internal circuitry consists of a resistor chain and a group of multiplexers that picks one of many nodes. It is important to note that this 'selector' also has a resistance in the order of 70 ohms. This is an important element in designing with such devices. A wiper current of 10mA will create a 700mV drop across the wiper contact! Therefore the wiper contact should only feed into high impedance nodes such as opamps to limit the current through this path. Any current pulled out from the wiper, or pushed into the wiper will cause a voltage differential across Rw (see figure below) that can cause a substantial error.

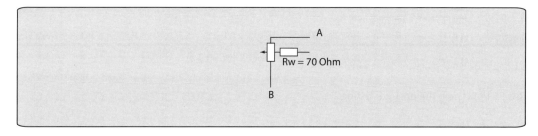

As current increases, the error increases as well. This is something that needs to be taken into account when designing with this kind of potentiometer.

Another important factor is that signals cannot go above the supply level or below the ground level of the integrated circuit. This may require DC biasing if AC signals are to be attenuated.

Microchip has an excellent application note (AN1316) on the usage of these devices. This is important reading material if you are going to use these devices in amplifier circuits.

34.2. DEVICE ADDRESS

The table below gives the base address for the MC40 (D) 1x devices.

	MSB							LSB
MCP4017 MCP4018 MCP4019	0	1	0	1	1	1	1	R/W
MCP40D17 MCP40D18 MCP40D19	0	1	0	1	1	1	0	R/W
MCP40D18A	0	1	1	1	1	1	0	R/W

This device has no address selection pins. The MCP40D18 is available with a different sub address by specifying a different order code.

34.3. USAGE

These devices have a single internal register that allows you to perform a number of operations. These devices are volatile and will lose their settings on a power cycle, or if the supply voltage dips below 1.5 volt.

The value is selected using a 7-bit word aligned on LSB.

The MCP401x devices are very simple in operation: address the device in read or write mode and read or write the contents as a byte.

34.3.1. Writing the potentiometer

Byte	Direction	D7	D6	D5	D4	D3	D2	D1	D0	CONTENT
0	OUT	0	1	1	1	1	1	*1	0	ADDRESS
1	OUT	0	P6	P5	P4	P3	P2	P1	P0	VALUE

*1 depends on the actual device installed

Bits P6 to P0 hold the desired setting.

34.3.2. Reading

Byte	Direction	D7	D6	D5	D4	D3	D2	D1	D0	CONTENT
0	OUT	0	1	1	1	1	1	*1	1	ADDRESS
1	IN	0	P6	P5	P4	P3	P2	P1	P0	VALUE

*1 depends on the actual device installed

Bits P6 to P0 hold the current setting.

34.4. THE 'D' VERSION

The MCP40D1x are intended for operation as SM bus and require the sending of an additional command byte. For every write operation two bytes need to be sent. The first byte needs to be all zeros as per SM-bus spec.

Byte	Direction	D7	D6	D5	D4	D3	D2	D1	D0	CONTENT
0	OUT	0	1	1	1	1	1	*1	0	ADDRESS
1	OUT	0	0	0	0	0	0	0	0	COMMAND
2	OUT	0	P6	P5	P4	P3	P2	P1	P0	VALUE

*1 depends on the actual device installed

Bits P6 to P0 hold the desired setting.

34.4.1. Reading

Reading these devices is more complicated as this involves a repeat start

Byte	Direction	D7	D6	D5	D4	D3	D2	D1	D0	CONTENT
0	OUT	0	1	1	1	1	1	*1	0	ADDRESS
1	OUT	0	0	0	0	0	0	0	0	COMMAND
										Repeat START
0	OUT	0	1	1	1	1	1	*1	1	ADDRESS
2	OUT	0	P6	P5	P4	P3	P2	P1	P0	VALUE

*1 depends on the actual device installed

Bits P6 to P0 hold the current setting.

34.5. ASSEMBLY DRAWING

The image below gives an overview of parts placements. All parts are located on the front of the board.

The devices come in very small SC 70 package. Be careful when placing these tiny devices.

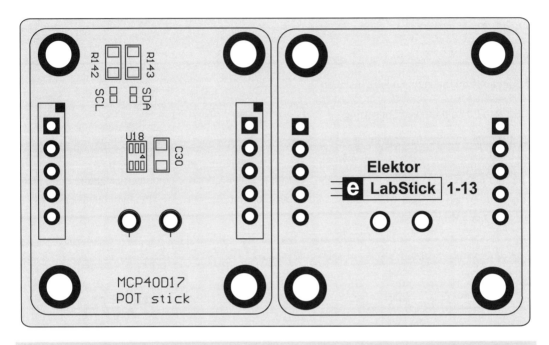

34.6. BILL OF MATERIALS

R142, R143	10 K Ohm SMD 0805
C30	100nF 25V X7R ceramic capacitor SMD 0805
U18	MCP4017, MCP4018, MCP4019, MCP40D17, MCP40D18 or MCP40D19 in SC 70 package
K44, K46	100 mils (2.54mm) pins. Strip of 5 pins.

35. LABSTICK 1-14: PCA9544 I2C MULTIPLEXER

The PCA9544 is a 4 channel multiplexer specifically designed for I2C busses. Although you can build your own I2C multiplexing system using an analog switch like a CD4052 or DG409 type circuit, the PCA9544 has advantages. First of all, the switch itself is addressable as an I2C device; this means you don't need to sacrifice valuable I/O pins to control the multiplexer.

The PCA9544 goes beyond simple switching, it has the capability to level shift each of its output channels (although output is a misnomer here, since they are actually bidirectional).

The device also features a pass through for an interrupt pin, so if you are using any devices that have an interrupt output, like the PCF8574 or PCF8563 for instance, these signals can also be passed through, including the level shifting capability.

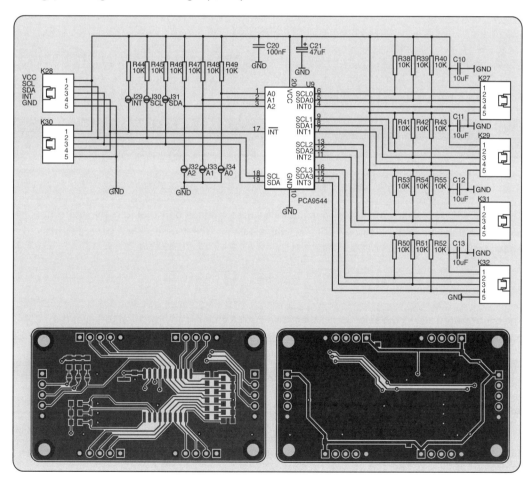

The LabStick replicates the familiar I2C connector 5 times. As usual the left and right (K28 and K30) connectors form the normal I2C bus, whereas the side connectors (K27, K29, K31, K32) now host the additional multiplexed busses.

As usual, the board provides the pull up resistors on the common I2C side, the interrupt pin and the address selection lines. A number of solder bridges allow switching these elements in the circuit, and selecting the device address for the PCA9544.

Note: the PCA9544 does not provide any kind of buffering, so it is very important that the total capacitance in a channel does not surpass the maximum allowed value for the bus. You can perform level shifting if you remove the pull up resistors for that channel from the board and provide them on your own circuitry tied to your supply voltage. Each channel can have different levels if so desired.

It is important to provide pull up resistors on all bus signals of the multiplexer including the INT pins. Failing to do so will block the internal logic in the PCA9544 and the device will not initialize correctly.

35.1. DEVICE ADDRESS

The PCA9544 has three address bits that allow up to 8 devices to be connected to the bus. Please note that it is not advisable to cascade these devices further.

	MSB							LSB
ADDRESS	1	1	1	0	A2	A1	A0	R/W

35.2. USAGE

The PCA9544 powers up with all four channels in isolation mode. It only monitors the interrupt pins and passes the state of those through. The master device must first send a control word to the PCA9544 to initialize it.

35.2.1. Control word

The control word allows the selection of the active output channel and the monitoring of the interrupt state.

	MSB							LSB
Control Byte	INT3	INT2	INT1	INT0	X	EN	S1	S0

The EN bit allows you to enable or disable the I2C multiplexer. After a power up this bit is set to zero, setting the bit to 1 enables the multiplexer operation.

The target channel is selected using bits s1 and s0.

The INT3 to INT0 bits indicate that an interrupt was received on any one of the channels. These bits give a real time representation of what is happening on the interrupt pins.

35.2.2. Interrupt logic

The interrupt logic monitors the incoming 4 interrupt lines. As soon as one goes low, the interrupt output will also go low. This feat is accomplished using 4 simple input NAND gates driving an open collector output.

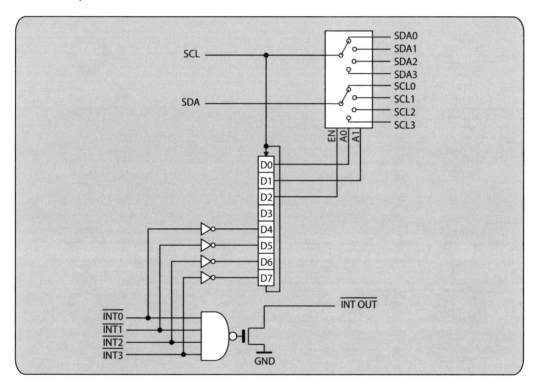

The incoming interrupt signals are each sent through an inverter and then displayed in the control register. Even though you can read and write to the control register, the INT bits are essentially read only.

In a system that is fully 'interrupt capable' the operation is very simple. If an interrupt is received from the PCA9544 you need to read the control word. The status of the top nibble will tell you which I2C channel is asserting the interrupt.

You can then select that channel by writing to the control register's bits c1 and c0. After interrogating the devices on that channel, the interrupt for that channel should clear. If this was the only interrupt source then the interrupt output pin of the PCA9544 will return to a logic high state.

Should another channel have an interrupt pending, the line will remain low. The master will now reread the control word and handle the next channel. For this mechanism to work the masters interrupt line must be level sensitive, or the interrupt handler must be written in such a way that it rechecks the control register for any additional pending interrupts prior to exiting.

The PCA9544 will perform the actual switch of the bus only when a STOP condition has been received. This is done explicitly so that the switchover only happens with SDA and SCL in the high state to avoid spurious glitches on the bus.

35.3. ASSEMBLY DRAWING

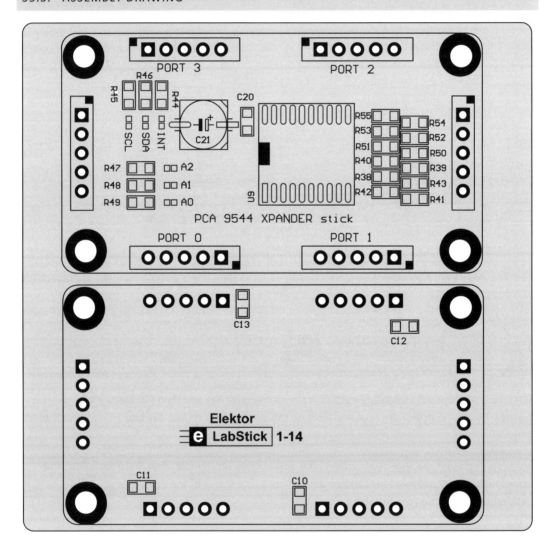

35.4. BILL OF MATERIALS

R38, R39, R40, R41, R42, R43, R44, R45, R46, R47, R48, R49, R50, R51, R52, R53, R54, R55, C20	10 K Ohm SMD 0805
	100nF 25V X7R ceramic capacitor SMD 0805
C10, C11, C12, C13	10uF 10V X5R OR X7R SMD 0805
C21	47uF 16V
U9	PCA9544
K27, K28, K29, K30, K31, K32	100 mils (2.54mm) pins. Strip of 5 pins.

36. LABSTICK 1-15: PCF8574(A) / PCA9XXX UNIVERSAL I/O EXPANDER

The basic I/O circuits on I2C were traditionally the PCF8574 and later the PCF8574A. A 16-bit version is also available under the PCF8575 devices.

The PCF8574 has open drain outputs that have a very weak pull up current source (less than 300 microampere). The outputs are capable of sinking up to 30 milliamperes.

An output pin that is programmed as logic 1 can be pulled low externally. This allows the pin to be used as an input (active low input).

The PCF8574 has a compare register. Whenever the programmed state on the outputs changes, because somebody is pulling a pin low, or they release a pin back to high, an interrupt will be given. This can trigger a master device so that it can scan the I2C bus for the attached PCF8574's and read the pin state.

The stick is universal and can also host other I/O expanders that come in the same type of package. There is a whole range of other devices out there that can be accommodated by this LabStick.

device	description	Base address
PCA9500	8-bit I/O PORT AND 2K EEPROM. Close INT jumper	0100xxx *1
PCA9534	PUSH / PULL output	0100xxx
PCA9538	Reset input instead of A2. Do not close A2 jumper onboard.	11100xx *2
PCA9554	Push / Pull output and pull up's.	0100xxx
PCA9554A	Push / Pull output and pull up's.	0111xxx
PCA9670	Fast mode + and reset. Close INT jumper (INT is replaced with reset)	0100xxx *3 *5
PCA9672	Fast mode +, reset and int. Do not close A2 onboard	01000xx *4
PCA9674	Fast mode +	0100xxx *5
PCA9674A	Fast mode +	0111xxx *5
PCF8574	Original port expander	0100xxx
PCF8574A	Original port expander alternate address	0111xxx
MAX7328	Enhanced 8574	0100xxx
MAX7329	Enhanced 8574A	0111xxx

*1: This device replaces the INT pin with Write protect for the EEPROM. The INT jumper must be closed to allow the EEPROM to be written.

*2: the A3 Pin is replaced by a Reset input. Only 4 devices are possible on the bus. Leave A3 on the board open. If you close it the device will be held in reset.

*3: the INT pin is replaced by a reset input. Make sure the INT pull up is installed by closing the jumper.

*4: this device can have 16 slave addresses. It behaves as the PCF8574 for standard operations on A0 and A1. If A0 and A1 are tied to SDA and SCL, other addresses can be formed. The datasheet has full details.

*5: this device allows for a whopping 126 slave addresses! By default it uses the PCF8574 like 0100xxx address. And it will function correctly on the board. By tying the A0 A1 and A2 pins to varying combinations of SCL and SDA the other combinations can be made. The datasheet has the full table of combinations to form the 128 slave addresses. Only the General call address and the Device ID address are not available. Care must be taken not to impede on any other addresses, for example like the 10-bit mechanism, HS mode master codes etcetera, if you use those in your system. If none of these special features are used, you can allocate them to these devices. If you only use VDD and VSS then the base addresses are as per PCF8574 or PCF8574A.

36.1. SCHEMATIC

Let's take a look at the schematic. The electronics are fairly straightforward. The classic pull ups on SDA and SCL can be turned on by closing jumpers J37 and J36 with a small dot of solder. The interrupt output can be provided with its own pull up resistor if you close jumper J35.

The back of the board has room for additional pull up resistors on the outputs should you desire these (R62 to R69).

Depending on which I/O expander you install you may not want to install these additional pull up resistors. For more information please read the remarks given in the previous table or consult the datasheet of the exact part you are going to install.

Jumpers J38 J39 and J40 allow you to adjust the slave address for devices that do have this capability. A closed jumper applies a logic zero while an open jumper applies logic 1. For these jumpers please consult the table above and/or datasheet of the specific device you are going to install. Some devices use one or more of the address pins for other purposes and you may not want to close the jumper on those parts.

The board has the standard 5 pin I2C connector that is common on these LabSticks.

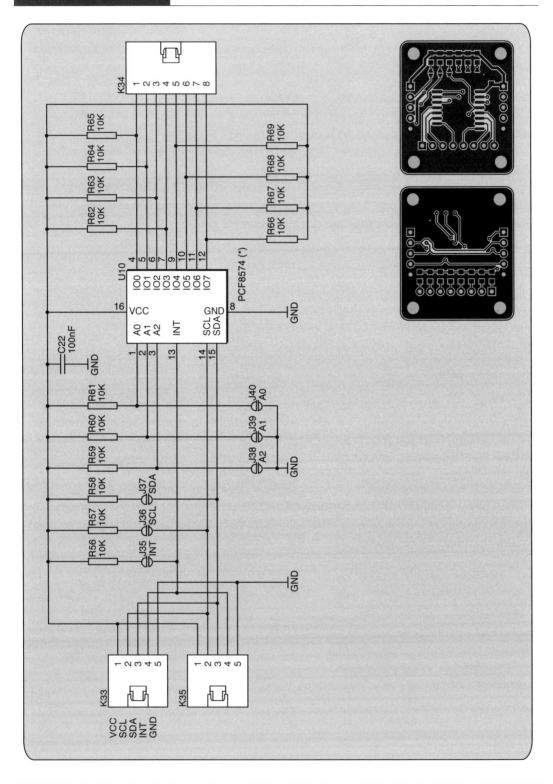

36.2. DEVICE ADDRESS

The device addresses have been covered in the previous table. However, for clarity I will repeat the two base addresses for both the standard and the 'A' version of the common parts. Any of the devices in that table can be positioned on these addresses with the exception of the PCA9672 which only uses 2 of the address bits.

36.2.1. PCF8574

ADDRESS	MSB							LSB
	0	1	0	0	A2	A1	A0	R/W

36.2.2. PCF8574A

ADDRESS	MSB							LSB
	0	1	1	1	A2	A1	A0	R/W

36.3. USAGE

Depending on the selected part the operations can be a bit different. I will categorize the operation according to the type of device

36.3.1. PCF8574 / PCF8574A / MAX7328 / MAX7329

These devices have a single register that is read- writable. If multiple writes are performed in one I2C operation, the device will update its output for every byte received from the master. If multiple reads are performed, the device will resample its I/O for every transmitted byte.

The byte received by the device is placed on its I/O pins. A logic zero will make the output pin go to hard ground. Note that you should not surpass the maximum IoL (Current Output Low) of the device you are using, as allowing excess current may damage the device. A logic one will return the pin to weak logic high. The maximum output is limited by the internal current source to 100 microampere. When driving CMOS or TTL circuitry this is sufficient. When driving external components that require more current it will be necessary to add external pull-up resistors.

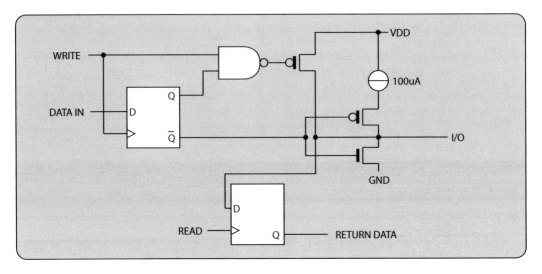

A read operation will return the current state on the I/O pins. If an external device is pulling the pin low, the read will return a logic 0. Any pin that has a sufficiently high level will return a logic 1.

Essentially when using these devices: simply write logic 1 to any pin you want to use as an input. Connect circuitry in such a way that it can pull the pin low. Never actively force a logic one onto a pin since this can cause short circuits when the output is activated.

36.3.1.1.1. WRITE OPERATION

Byte	Direction	D7	D6	D5	D4	D3	D2	D1	D0	CONTENT
0	OUT	0	1	0	0	A2	A1	A0	0	ADDRESS
1	OUT	X	X	X	X	X	X	X	X	DATA

36.3.1.1.2. READ OPERATION

Byte	Direction	D7	D6	D5	D4	D3	D2	D1	D0	CONTENT
0	OUT	0	1	0	0	A2	A1	A0	1	ADDRESS
1	IN	X	X	X	X	X	X	X	X	DATA

It is allowed to read or write multiple bytes in a sequence. Each byte sent in write mode will be applied in the same sequence to the output. When reading, every byte read will re-sample the state of the input pins.

36.3.1.1.3. INTERRUPT

The interrupt pin will activate whenever the state of the I/O pins change.

36.3.2. PCA9534

These devices are pin compatible upgrades of the classic PCF8574, though they function differently internally and have a different I/O structure.

The PCA9534 has a push pull output that can sink 20mA and source up to 50mA. There is no internal pull up. A pin that is programmed as input must have an external pull up mechanism or have push-pull driving capability itself.

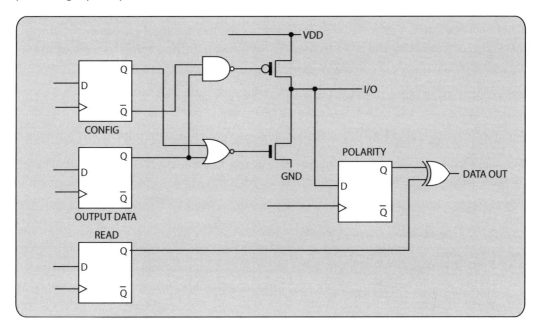

Register wise the PCA9534 is different. This device has 4 internal registers that are selected by sending a command word.

36.3.2.1.1. COMMAND WORD

The command word acts as the selector for the register we want to perform an operation on.

Bit	7 (MSB)	6	5	4	3	2	1	0 (LSB)
ADDRESS	0	0	0	0	0	0	R1	R0

The bits R1 and R1 select one of a possible 4 registers in the device

Register	Mode	R1	R0
Input register	R	0	0
Output register	R/W	0	1
Polarity inversion register	R/W	1	0
Configuration register	R/W	1	1

The input register is read only, so writing to this register has no effect.

36.3.2.1.2.　INPUT REGISTER

The input register returns the current state of the I/O pins with the inversion register applied.

36.3.2.1.3.　INVERSION REGISTER

This register allows you to control the level sensitivity of the input. A logic 0 returns the state of the pin as is, while a logic 1 returns the inverted state of the pin. This is useful if you don't want to bother decoding active low and active high signals. Simply program the correct polarity mask and the device will return logic 1 for an active input.

In the case of the polarity register you can read 0 as O for original and 1 as I for Inverted.

36.3.2.1.4.　OUTPUT REGISTER

The output register contains the data that needs to be sent to the output. This data is masked by the configuration register before it is sent out. Data for a pin configured as input will not be put on the bus.

36.3.2.1.5.　CONFIGURATION REGISTER

The configuration register allows you to define which pins are input and which are output. A logic 0 sets a pin as output (O for output) while a logic 1 sets the pin as input (I for input). After power up all pins are defined as inputs.

36.3.2.1.6.　INTERRUPT PIN

The interrupt pin will activate if the state of one of the input pins changes. You will not get an interrupt by writing to the output register. There is a possibility for a false interrupt if you change the configuration register. If the pin was set to low in output mode and is pulled high externally then switching the pin to input will trigger an interrupt since the state now has changed from a self applied 0 to an externally applied 1.

36.3.2.2.　BUS TRANSACTIONS

Every write transaction must specify the command byte to select the register to be written to. The register pointer does not auto increment. Once set it remains. Consecutive writes without a stop condition will simply be written to the same selected register.

36.3.2.2.1. WRITE TRANSACTION

Byte	Direction	D7	D6	D5	D4	D3	D2	D1	D0	CONTENT
0	OUT	0	1	0	0	A2	A1	A0	0	ADDRESS
1	OUT	0	0	0	0	0	0	R1	R2	REGISTER
3	OUT	x	X	X	x	x	x	x	x	DATA

Multiple bytes can be written in one transaction. The register selector does not auto increment so the data will end up in the selected register. This allows you to stream data to, or from, the output and input ports.

36.3.2.2.2. READ TRANSACTION

A read operation will be performed on the last register accessed in write mode. It is possible to issue a repeat start to perform this feature, or you can simply stop the write transaction and re-access the device in read mode.

Byte	Direction	D7	D6	D5	D4	D3	D2	D1	D0	CONTENT
0	OUT	0	1	0	0	A2	A1	A0	1	ADDRESS
1	IN	x	x	x	x	X	x	x	x	DATA

36.3.2.2.3. READ TRANSACTION WITH REPEAT START

Byte	Direction	D7	D6	D5	D4	D3	D2	D1	D0	CONTENT
0	OUT	0	1	0	0	A2	A1	A0	0	ADDRESS
1	OUT	0	0	0	0	0	0	R1	R0	REGISTER
RS										REPEAT START
0	OUT	0	1	0	0	A2	A1	A0	1	ADDRESS
1	IN	x	x	x	x	x	x	x	x	DATA

36.3.3. PCA9538

The PCA9538 is largely identical to the PCA9534 I described earlier. The register structure and its operation is identical to the PCA9534. The I/O structure and specifications are also identical to those of the PCA9534. Only the base address is different. Due to the replacement of pin A2 with a reset input there are less base addresses.

36.3.3.1.1. BASE ADDRESS

Bit	7 (MSB)	6	5	4	3	2	1	0 (LSB)
ADDRESS	1	1	1	0	0	A1	A0	R/W

36.3.3.2. USAGE

For the usage of the device please refer to the description of the PAC9534. Remember to correct the base address of the device since they are very different.

36.3.4. PCA9554 / PCA9554A

These devices are almost identical in operation to the PCA9534, which has been already covered. The PCA9554 is also available with an A suffix to indicate the different base address, just like the PCF8574A. From a software perspective, the device behaves exactly the same as a PCA9534.

Hardware wise the device adds an additional pull up resistor in parallel with the top transistor. This resistor is in the order of 100 K. ohms. When the pin is defined as input this removes the requirement for an external pull up. When configured as an output this additional resistor has no purpose since the I/O driver is push pull anyway.

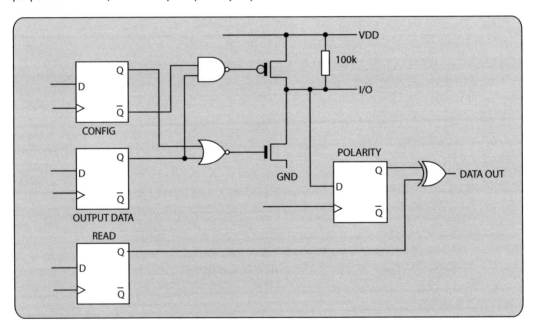

36.3.5. PCA9670

These devices are identical in I/O structure and register level operations to the classic PCF8574. The interrupt output is replaced by a reset input.

The device has several additional enhancements such as a rapid pull up mechanism on the output. During a write operation an additional transistor is briefly activated to help charging the possible capacitive loads on the output pin.

On the I2C side the device supports fast mode+ up to 1MHz and it has improved bus drive strength to accommodate the higher speeds and greater bus loading, also the device has an impressive range of slave addresses that are set by tying the address lines not only to VDD or VSS but also to SDA or SCL. During power up the device tests the electrical connection and determines its address. The table is too large to reproduce here, so please consult the correct datasheet for this device to find out all the possible combinations.

This device also supports advanced I2C operations such as device id and the software reset call.

36.3.6. PCA9672

The PCA9672 is a variation on the PCA9670 that replaces the A2 pin with the reset input and restores the interrupt output. The same as the PCA9670, the address pins A1 and A0 (A2 does not exist in this device) can be tied to SDA and SCL to create additional addresses. The table is substantially shorter than that of the PCA9670. Please consult the actual datasheet for more information.

36.3.7. PCA9674 / PCA9674A

The PCA9574 and PCA9574A are both drop in replacements for the PCF8574 and PCF8574A. They support Fast mode+ at clock rates of up to 1MHz and have higher drive strength for SDA and SCL to compensate for possible capacitive loading.

These devices allow for a wider range of addresses. If the A0, A1 and A2 lines are connected in the same manner as with the PCF8574 and PCF8574A, the addressing is identical, however, you can also connect with SDA and SCL. These additional combinations allow for 126 possible slave addresses. (Only addresses 0000000 and 1111100 are disallowed since these are the general call and device ID addresses.)

The operation, as well as the I/O structure, is identical to that of the PCF8574.

36.3.8. PCA9500

The PCA9500 is essentially a PCF8574 combined with an EEprom, all in one device. This device is unique in the I2C compatible device list because it responds to two I2C addresses. It reacts to the standard PCF8574 address range (0100xxx) for the I/O port as well as the base EEprom address 1010xxx.

The I/O capabilities are both in function and register compatibility identical with the PCF8574.

The EEprom part behaves as a standard PCF8582, so I will not go into detail on the operation of this part. There is a dedicated LabStick for EEprom and the operation is explained in full there.

36.4. ASSEMBLY DRAWING

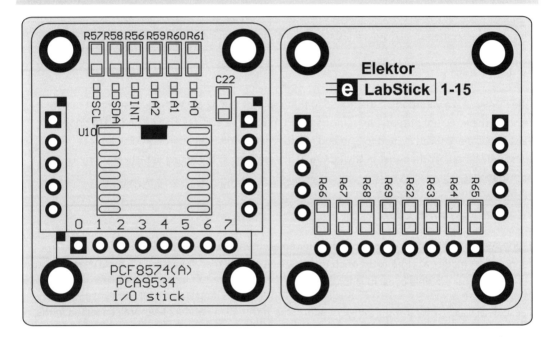

36.5. BILL OF MATERIALS

R56, R57, R58, R59, R60, R61, R62, R63, R64, R65, R66, R67, R68, R69	10 K Ohm 0805 SMD
C22	100nF 25V X7R ceramic capacitor SMD 0805
U10	PCF8574 / PCF8574A / or similar in SO16 wide body.
K33, K35	100 mils (2.54mm) pins. Strip of 5 pins.

37. LABSTICK 1-16: SAA1064 7 SEGMENT DISPLAY

The display controller allows the operation of 4 x 7 segment LED displays. The board has the through-hole displays mounted on the front and the SAA1064 is mounted on the back, together with all the required support electronics.

The SAA1064 is one of the oldest I2C devices around and is constructed in a bipolar process, as opposed to newer CMOS circuitry.

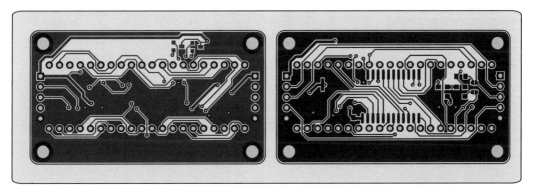

A look at the schematics on the next page doesn't bring any real surprises. Resistors R27 and R28 together with jumpers J21 and J22 allow you to switch in the pull ups for both SDA and SCL.

The device sub addressing mechanism is non-standard. This device only has a single address pin but allows for 4 possible addresses. If you tie the pin to either ground or power you have two of the possible addresses. Applying a voltage equal to roughly 3/8 VCC or 5/8 VCC gives you the other two possible addresses.

R30	R31	MSB							LSB
OPEN	1K	0	1	1	1	0	0	0	R/W
1K5	1K	0	1	1	1	0	0	1	R/W
1K	1K5	0	1	1	1	0	1	0	R/W
1K	OPEN	0	1	1	1	0	1	1	R/W

Capacitor C18 sets the oscillator frequency for the internal clock generator. This clock controls the internal multiplexing of the display logic. The displays are duplexed using transistors Q1 and Q2. The SAA1064 has current mode drivers that control the LED current under software control so no series resistors are required for the displays.

LabStick 1-16: SAA1064 7 segment display

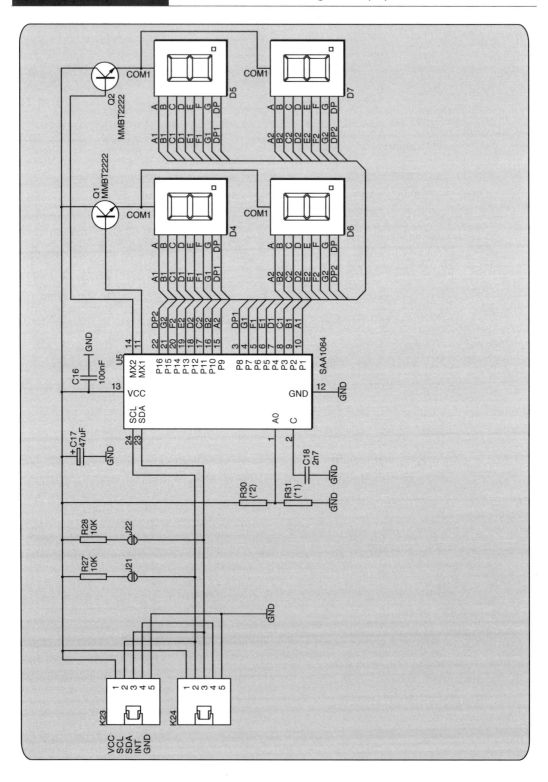

37.1. USAGE

The SAA1064 is not symmetrical for read/write behaviour. Only one register can be read and that is the status register and it contains only a power flag.

37.1.1. Status register (READ ONLY)

STATUS Byte	MSB							LSB
	PR	0	0	0	0	0	0	0

The PR bit indicates that a power cycle has occurred since the last access. This can be used as an indicator to the controlling master that the display contents need refreshing.

In write mode the SAA1064 has a number of registers, when addressing, the first byte that needs to be transmitted is the register selector.

37.1.2. Register select

STATUS Byte	MSB							LSB
	0	0	0	0	0	RS2	RS1	Rs0

The possible combinations for RS2, RS1 and RS0 are:

Register	R2	R1	R0	Description
CONTROL	0	0	0	Controls device operation and LED current.
DIGIT 1	0	0	1	Controls segments of digit 1
DIGIT 2	0	1	0	Controls segments of digit 2
DIGIT 3	0	1	1	Controls segments of digit 3
DIGIT 4	1	0	0	Controls segments of digit 4

After sending the register select byte, you can continue by filling with data. Each byte received will auto increment the pointer to the internal register. This pointer will roll over from code 111 to 000. So you can write bursts of data in a continuous stream. Just remember to insert 3 zero bytes to roll over and back to register 0 (control register).

37.1.3. Control register

STATUS Byte	MSB							LSB
	0	I2	I1	I0	TEST	B1	B0	SD

The Static Dynamic bit (Bit 0: SD) switches the device operation from 2 digit to 4 digit. In static mode (SD=0) only digits 1 and 2 are in operation. No multiplexing is performed. When the SD bit is set then all 4 digits are in operation.

The blanking bits B1 and B0 control the blanking of a group of digits. When B1 is set to 0 then digits 1 and 3 are blanked. When B1 is set to 1 then digits 1 and 3 behave normally by displaying the contents of their digit registers.

Likewise, the B0 bit controls the operation of digits 2 and 4.

The TEST bit allows you to turn on all digits and all segments to perform a lamp test for example.

Bits I2, I1 and I0 form a 3-bit DAC to control the current through the LED's. The LED current is equal to the DAC setting x 3mA. A DAC setting of 000 turns all LED's off, where a setting of 111 (7 decimal) sends of course 21 mA through the segments.

37.1.4. Digit registers

The digit registers allow the manipulation of the segments for each individual digit.

	MSB							LSB
DIGIT Byte	D7	D6	D5	D4	D3	D2	D1	DD
SEGMENT assignment	DP	G	F	E	D	C	B	A

There is no fixed wiring and it is up to the designer/builder to figure out which bit he wants to use for what segment. I use the normal segment order and used bit D7 to control the decimal point.

The next table shows the segment assignment and a number of combinations to form the most commonly used digits.

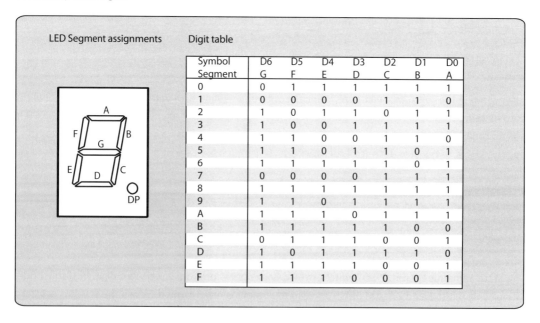

LED Segment assignments

Digit table

Symbol / Segment	D6 G	D5 F	D4 E	D3 D	D2 C	D1 B	D0 A
0	0	1	1	1	1	1	1
1	0	0	0	0	1	1	0
2	1	0	1	1	0	1	1
3	1	0	0	1	1	1	1
4	1	1	0	0	1	1	0
5	1	1	0	1	1	0	1
6	1	1	1	1	1	0	1
7	0	0	0	0	1	1	1
8	1	1	1	1	1	1	1
9	1	1	0	1	1	1	1
A	1	1	1	0	1	1	1
B	1	1	1	1	1	0	0
C	0	1	1	1	0	0	1
D	1	0	1	1	1	1	0
E	1	1	1	1	0	0	1
F	1	1	1	0	0	0	1

37.2. ASSEMBLY DRAWING

This board is a bit trickier to assemble than your average LabStick. Begin by placing the IC followed by the passive components on the back of the board. Figure out what combination you want to use for R30 and R31. Do not install C17 yet as it will interfere with the next job. Install the headers on the back of the board.

Flip the board over and solder the headers, snip off the excess length of the pins sticking out the front of the board. Now install the two transistors on the front and populate the 4 displays, then turn the board back over. Once the displays are soldered in you can then install C17.

If you want to install the displays in pin headers that is fine, and if you snip off the pins of the pin header flat with the board, they should not interfere with the display placement.

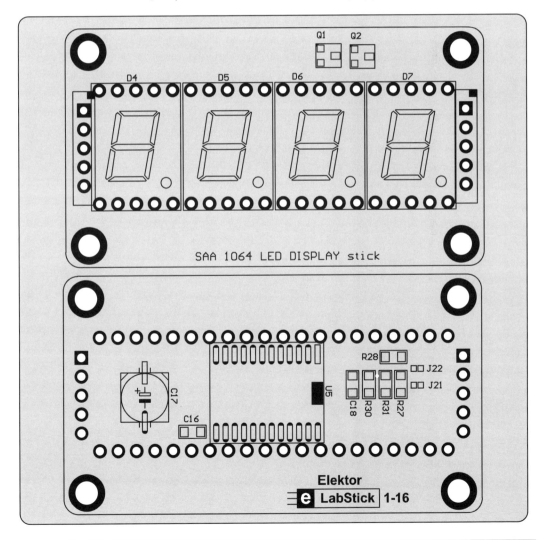

37.3. BILL OF MATERIALS

R27, R28	10 K Ohm SMD 0805
R30, R31	Depends on address. See description under address.
C16	100nF 25V X7R ceramic capacitor SMD 0805
C17	47uF 16V SMD Electrolytic
U5	SAA1064 in SMD
D4, D5, D6, D7	7 Segment Common ANODE display
Q1, Q2	MMBT2222 or BC847 PNP transistor in SOT-23
K23, K24	100 mils (2.54mm) pins. Strip of 5 pins.

38. APPENDIX 1

38.1. PART REFERENCES

In the various 'bills of materials' throughout this book, only simple part descriptions have been given. Most of you will be able to find these parts without problems. However, some of the parts like pushbuttons, displays etcetera are not always easy to cross reference without additional information.

That's what this section is about. I will give you the order codes from Digikey and/or Mouser. Using these part numbers and the information available on Digikey's or Mouser's website you can easily cross reference the parts to your favourite supplier.

38.2. PASSIVE PARTS

Resistors

Component	Description	Digikey Order Code.
22r	22 Ohm resistor 5% 0805 SMD	RMCF0805JT22R0CT-ND
100r	100 Ohm resistor 5% 0805 SMD	RMCF0805JT100RCT-ND
150r	150 Ohm resistor 5% 0805 SMD	RMCF0805JT150RCT-ND
220r	220 Ohm resistor 5% 0805 SMD	RMCF0805JT220RCT-ND
330r	330 Ohm resistor 5% 0805 SMD	RMCF0805JT330RCT-ND
470r	470 Ohm resistor 5% 0805 SMD	RMCF0805JT470RCT-ND
1K	1K Ohm resistor 5% 0805 SMD	RMCF0805JT1K00CT-ND
10K	10K Ohm resistor 5% 0805 SMD	RMCF0805JT10K0CT-ND

Capacitors

Component	Description	Digikey Order Code.
100nF	Capacitor Ceramic, 100nF (0.1uF), X5R or X7R, 0805	PCC1812CT-ND
47uF	Electrolytic Capacitor, 47uF 16V, D package	PCE4000CT-ND
22pF	Ceramic Capacitor, 22pF, NP0 or C0G, 0805	478-1308-1-ND
0F47	Capacitor Goldcap, 0.47F, 5.5V	P6974-ND
5..25pF	Trim Capacitor Murata TZY series, 5 to 25pF adjustable	490-2004-2-ND
1000uF 25V	Electrolytic Capacitor, 1000uF, 25V, 5mm pitch (100mil)	P13121-ND
220uF 16V	Electrolytic Capacitor, 220uF, 16V, 5mm pitch (100mil)	P10370TB-ND
470pF	470pF, NP0 or C0G, 0805	445-4134-1-ND
2n7	Ceramic Capacitor, 2700 pF (2n7), 0805	PCC272BNCT-ND
10uF	Ceramic Capacitor, 10uF, 10V or 16V, X5R or X7R, 0805	490-1709-1-ND

Connectors

Component	Description	Digikey Order Code.
USB B Type	USB B Type Right Angle Connector	AE1085-ND
Terminal block	Terminal block, 10 contact, 3.81mm pitch, wire to board	609-3926-ND
Pin header	Pin header 100mil (2.54mm). These can be cut to the desired pin count.	WM6436-ND
Test points	Keystone 5010 Test Point	5010K-ND
Dc power jack	DC Power jack Right Angle	SC1153-ND
Pushbutton 6mm	Tactile switch Thru hole 6mm.	SW405-ND
Cap for pushbutton	Assorted colours	Digikey : (square cap) SW845-ND – Orange SW250-ND - Blue SW249-ND - Yellow SW252-ND - Red SW251-ND - White SW450-ND - Light Green SW451-ND – Black Mouser : (round cap) 101-0200-EV (Black), 101-0202-EV (Red), 101-0204-EV (Yellow), 101-0205-EV (Green), 101-0206-EV (Blue), 101-0209-EV (Ivory)

Crystals

Component	Description	Digikey Order Code.
Crystal 20 MHz	Crystal 20MHz HC49/U or HC49/US body	X439-ND
Crystal 32.768 KHz	Crystal 32.768Khz. round body	535-9032-ND

38.3. SEMICONDUCTORS

Discrete Circuitry

Component	Description	Digikey Order Code.
BAV99	Dual diode in SOT23 package	BAV99WT1GFSCT-ND
1SMB5916	Zener Diode, 4V3, SMB (DO214AA) package	1SMB5916BT3GOSCT-ND
BAT54	Single diode BAT54 SOT-23 Careful! The suffix denotes single or multi diode and the orientation in the package. Only the BAT54 without suffixes or BAT54FILM will work.	497-7162-1-ND
TOPLED Red	Red LED, PLCC-2 (TOPLED)	475-1180-1-ND
TOPLED Orange	Orange LED, PLCC-2 (TOPLED)	475-2745-1-ND
TOPLED yellow	Yellow LED, PLCC-2 (TOPLED)	475-2562-1-ND
TOPLED Blue	Blue LED, PLCC-2 (TOPLED)	475-1387-1-ND
Toppled green	Green LED, PLCC-2 (TOPLED)	475-2679-1-ND
1n4007	DIODE 1A 1000V DO41	1N4007FSCT-ND
MMBT2222	TRANS NPN 40V 350MW SMD SOT23	MMBT2222A-FDICT-ND
BAS16	Diode 75V 200mA SOT23	BAS16-FDICT-ND
BAS116	Diode 75V 200mA SOT23 LOW LEAKAGE	BAS116-FDICT-ND
7 Segment LED	LED display common anode	160-1575-5-ND

Integrated circuits

Component	Description	Digikey Order Code.
74HCT245 or 74HC245	Octal transceiver in 20 pin SOP (5.3mm wide!) Fairchild part number MM74HC245ASJ or MM74HCT245SJ. Only the 5.3mm version will fit. Texas Instruments and Fairchild are the only manufacturers making this package.	Mouser : 512-MM74HC245ASJ 512-MM74HCT245SJ Digikey: MM74HC245ASJ-ND MM74HCT245SJ-ND
ULN2803	Octal Darlington array 18 pin SOIC wide body (7.15mm)	296-15777-2-ND
LCD	LCD 2 lines x 8 character. Top connector	NHD-0208AZ-RN-GBW-ND
74HCT138	3 to 8 decoder 16 pin SOIC (3.9mm)	497-1900-1-ND
PIC18F2550	8-bit PIC processor with USB. This device needs programming with I2CPROBE.BIN or you can order it pre-programmed from the elektor shop.	PIC18F2550-I/SO-ND
L7805	Voltage regulator 5 volts.	497-1441-5-ND

I2C Integrated Circuits

Component	Description	Digikey Order Code.
PCF8574	I/O expander 16pin SOIC wide body (7.15mm)	568-1077-1-ND
PCF8574A	I/O expander 16pin SOIC wide body (7.15mm) alternate address	568-1074-1-ND
PCA9534	8-bit I/O expander, 16pin SOIC wide body (7.15mm)	568-1833-1-ND
PCA9554A	8-bit I/O expander, 16pin SOIC wide body (7.15mm)	568-1051-5-ND
PCA9553	LED driver. 8 pin SOIC (3.9mm). Part numbers are for different sub address.	568-3390-5-ND 568-3391-5-ND
PCA9544	I2C Multiplexer, 20pin SOIC wide body (7.15mm)	568-1860-1-ND
MCP40D17	Potentiometer SC70 body, 10K	MCP40D17T 103E/LTCT-ND
ADC121	12-bit ADC SOT23-6	ADC121C027CIMKCT-ND
PCF8591	AD/DA combo 16 pin SOIC wide body (7.15mm)	568 1088 1-ND
SAA1064	4 Digit LED driver	568 1107 5-ND
MCP4725	12-bit DAC. Suffix determines sub address. SOT23 6 package.	MCP4725A0T E/CHCT-ND MCP4725A1T E/CHCT-ND MCP4725A2T E/CHCT-ND MCP4725A3T E/CHCT-ND
PCF8563	Real time clock 8 pin SOIC (3.9mm)	568-1068-1-ND
24C02	EEprom 8 pin SOIC (3.9mm)	497-8552-1-ND
LM75	Temperature monitor 8 pin SOIC (3.9mm)	568-4688-6-ND

39. APPENDIX 2

39.1. ADDITIONAL READING

There is lots of documentation out there that deals with the I2C bus. Most manufacturers' datasheets have detailed explanations on how to operate a particular device and the bus in general.

The gold standard for I2C remains the official I2C Specification by NXP Semiconductors (formerly Philips Semiconductors)

The document is revised regularly and its precise location on NXP's website tends to shift. The easiest way to find this document is to Google for 'I2C Spec'

40. INDEX

16-bit I/O Expander, 95
2408, 116
2416, 116
24c01, 116
24C01, 149
24C02, 116, 117, 149, 150, 151, 152, 154, 242
24C04, 116, 117, 149, 150
24C08, 149, 150
24C1024, 117, 149, 150
24C128, 117, 149
24C16, 117, 149, 150
24C256, 117, 149
24C32, 117, 149, 150, 152
24C512, 117, 149, 150
24C64, 117, 149
2N7002, 67
400 Kbit/s, 47, 53
4051, 65
4052, 65, 128
4053, 65
4066, 59, 65
5501, 63
5502, 63
74138, 162, 166, 167
74245, 193
74HC05, 83
74ol6000, 64
74xx4052, 65
7805, 145
8042, 90
8574, 222
A/D, 105
A/D converter, 43, 96, 112, 114, 202, 203, 205
abort, 70
ACCESS bus, 17
ACCESS Bus, 90
Access mode, 37
Accuracy, 111
ack, 70
ACK, 24, 26, 31, 32, 33, 37, 42, 43, 46, 69, 70, 76, 81, 136
acknowledge, 33
ACKNOWLEDGE, 25, 26, 31, 32, 33, 35, 36, 37, 44, 46, 54, 69, 72, 136
ACPI, 91
ACTIVE MASTER, 23
ADC, 209
ADC081, 202, 203, 204
ADC101, 202
ADC121, 129, 202, 203, 242

ADDRESS, 25, 29, 30, 35, 149, 151, 152, 165, 174, 178, 198, 200, 203, 204, 205, 207, 208, 214, 215, 218, 225, 226, 227, 229
Addressing, 29, 30
Adressing, 55
ADUM1250, 63, 64
ADUM1251, 63
ADUM2250, 63
ADUM2251, 63
Advanced Telecom Computing Architecture, 17
AN1316, 214
ANALOG to Digital converters, 108
ATCA, 52
Audio, 115
Avit Research, 71
Baseboard Management Controller (BMC), 52
Basic, 24, 83
bidirectional, 68
BRM DAC, 107
BSS138, 67
Bus Architecture, 22
BUS MASTER, 22
BUS SLAVES, 22
bus stall, 32
bus topology, 22
BUS Transaction, 23
Calibre, 71
CBUS, 44, 45, 53
Combined data format, 38
Configuration, 44
controller, 67
Corelis Buspro, 71
CY8C9560A, 98
Cypress Ez USB, 71
D/A, 105
DAC, 105, 106, 107, 108, 109, 111, 197, 198, 199, 200, 209, 210, 236, 242
DATA, 25, 31, 35, 45, 76, 77, 78, 79, 151, 152, 165, 200, 207, 208, 226, 229
debug, 73
Debugging, 80
DEVICE address, 30
Device identification, 47
DG409, 128
Digital Domestic Bus, 17
DIMM, 51, 90
display, 99
displays, 233
distance, 56, 57
DLEN, 46

Index LabWorX

DNL, 112
DS1337, 122
DS4520, 97
dual slope, 109
E2Prom technology, 119
e2proms, 152
E^2PROMs, 38
eeprom, 148
EEprom, 30, 32, 51, 79, 97, 117, 120, 121, 129, 148, 149, 150, 151, 152, 153, 154, 155, 197, 198, 231, 232, 242
EEPROM, 32, 38, 45, 79, 90, 97, 116, 119, 120, 155, 199, 200, 222
Enhanced I2C, 53
Ethernet, 52, 53, 56, 57
exceptions, 44
expanders, 125
extend, 59
Extended addressing, 54
Fast, 48
Fast mode, 47, 48, 53, 222, 231
FAST mode, 53, 54
Flash convertor, 108
FRAM, 120
General call, 44
GIVING ACK, 26
HD44780, 101, 162, 165
Hendon 5501, 63
High speed, 46, 48
High speed masters, 48, 49
I2C bridge, 49
I2C protocol, 17, 18, 22, 24, 42, 51, 68
I2CProbe, 132, 137
ICH 82801, 90
idle, 18, 22, 24, 26, 27, 35, 37, 39, 40, 162
IDLE, 26
IES5502, 63
INL, 113
Intelligent Platform Management Interface, 17, 52
Interrupt, 45
IPMI, 17, 52, 53
ISL1208, 121
isolate, 63
isolators, 122
Joint technology, 71
Jupiter Instruments, 71
LabStick, 132, 145, 148, 156, 162, 173, 177, 185, 190, 197, 202, 209, 212, 217, 218, 222, 232, 233, 237
LabSticks, 129, 132, 223
LCD, 99, 100, 101, 115, 129, 162, 164, 165, 172
LED, 64, 74, 98, 99, 101, 102, 103, 129, 134, 144, 147, 156, 157, 158, 159, 160, 161, 185, 189, 192, 195, 233, 235, 236, 242
Level shifting, 66

LM75, 51, 129, 173, 174, 175, 176, 242
LM75A, 173, 174, 175
lower clock frequency, 56
LTC4310, 124
LTC4311, 54, 57
MASTER, 23, 24, 26, 27, 28, 29, 31, 32, 33, 35, 36, 37, 38, 41, 42, 45, 93, 136, 139
master device, 22, 28, 218, 222
MASTER LISTENER, 26
MASTER TALKER, 23, 27
MASTERS, 31, 38, 45
MAX7328, 222, 225
MAX7329, 222, 225
MCP23009, 97
MCP23017, 95, 96
MCP23018, 95, 96, 97
MCP4017, 212, 213, 214, 216
MCP4018, 212
MCP4019, 212, 213, 214, 216
MCP401x devices, 212, 214
MCP40D1x, 212, 215
MCP4725, 129, 197, 198, 201, 242
Micro Computer Control, 71
Microsoft, 90, 91
monitor, 71
MOSFET, 65, 67, 119, 120, 159
MOSFETS, 67
multi byte, 35
multi master, 17, 19, 22, 24, 40, 42, 44, 45, 48, 51, 64, 68, 74, 82, 90
Multimaster, 74
MULTIMASTER, 40
multiplexers, 65, 125, 126
nACK, 26, 31, 32, 33, 37, 49, 69, 70, 126, 136, 138
noise, 19
NXP, 56, 57, 59, 61, 62, 66, 67, 71, 75, 82, 83, 90, 115, 125, 243
NXP9635, 102
Opcode, 45
open drain, 18
optocouplers, 62, 63, 64, 122, 124
P82B715, 60, 61, 66
P82B96, 57, 61, 62, 124
P87B715, 59
Page writing, 152
PB82B9, 64
PB82B96, 62, 63
PCA8550, 97
PCA8563, 178, 182
PCA8581, 149
PCA9500, 222, 231
PCA9516, 66
PCA9518, 66
PCA951x, 65

I2C Bus 245

PCA9531D, 75
PCA9532, 103
PCA9533, 156, 157, 160, 161
PCA9534, 185, 189, 222, 227, 229, 230, 242
PCA9536D, 75
PCA9538, 222, 229
PCA9538D, 75
PCA9540BD, 75
PCA9541D, 75
PCA9543AD, 75
PCA9544, 66, 129, 217, 218, 219, 220, 221, 242
PCA9551D, 75
PCA9553, 129, 156, 157, 159, 160, 161, 242
PCA9554, 129, 162, 164, 166, 167, 168, 169, 172, 222, 230
PCA9554A, 222, 230
PCA9564, 67
PCA9591, 210
PCA9600, 62, 64
PCA9670, 222, 230, 231
PCA9672, 222, 225, 231
PCA9674, 222, 231
PCA9674A, 222, 231
PCA9691, 108, 210
PCF85102, 149
PCF85103, 149
PCF85116, 75, 149
PCF8534, 100
PCF8563, 129, 177, 178, 180, 184, 217, 242
PCF8563TD, 75
PCF8573, 121
PCF8574, 37, 55, 65, 74, 79, 93, 94, 95, 98, 129, 169, 172, 185, 192, 195, 217, 222, 223, 225, 227, 230, 231, 232, 242
PCF8574A, 55, 93, 94, 129, 222, 223, 225, 230, 231, 232, 242
PCF8574TS, 75
PCF8575, 95, 222
PCF8581, 116
PCF8582, 79, 116, 148, 149, 150, 155, 232
PCF8583, 121
PCF8584, 67
PCF8591, 108, 129, 202, 209, 210, 211, 242
PCF8594, 149
PCF8598, 149
Perovskite, 120
PIC18F2550, 132, 144
PM bus, 50, 51, 52
PMBUS, 51
potentiometer, 212
probe, 132
protected input, 190
protected output, 185

protocol, 18, 24, 25, 47, 48, 50, 52, 53, 56, 67, 68, 69, 90, 126, 132
Protocol, 24
PseudoCode, 76
Pulse Width Modulation, 160
PUSH / PULL, 222
PWM, 107
R/W, 29, 30, 31, 35, 36, 37, 38, 39, 47, 54, 79, 137, 157, 165, 174, 178, 198, 203, 214, 218, 225, 227, 229, 233
R-2R network DAC, 106
READ address, 30
Reading, 36, 37
Real time clocks, 121
realtime clock, 177
reflections, 20
repeat start, 69
repeated start, 69
resolution, 111
RS232, 29, 72
RTC, 177
S x and S y lines, 59
SA56004ED, 75
SAA1064, 101, 129, 233, 235, 238, 242
SAA1300, 92
SAR, 108, 109
SC 70, 215, 216
Schmitt triggers, 47, 54
SCL, 18, 19, 22, 23, 26, 27, 28, 29, 31, 32, 33, 34, 36, 39, 40, 41, 42, 43, 45, 46, 47, 48, 49, 50, 51, 54, 56, 57, 58, 62, 64, 67, 68, 69, 72, 73, 74, 77, 78, 80, 81, 82, 83, 84, 92, 98, 123, 132, 134, 141, 142, 148, 156, 162, 173, 177, 185, 192, 197, 203, 209, 210, 220, 223, 231, 233
SDA, 18, 19, 22, 23, 26, 27, 28, 29, 31, 32, 33, 35, 36, 37, 39, 40, 41, 42, 43, 45, 46, 47, 48, 49, 50, 51, 54, 56, 57, 58, 62, 64, 67, 69, 70, 72, 73, 74, 77, 80, 81, 82, 83, 84, 92, 98, 123, 132, 134, 139, 141, 142, 148, 156, 162, 173, 177, 185, 192, 197, 203, 209, 210, 220, 223, 231, 233
Serial Clock (SCL), 18
Serial Data (SDA), 18
Sigma Delta, 111
single slope, 109
SLAVE, 23, 26, 32, 35, 36, 37, 38, 40, 47, 136
SLAVE LISTENER, 26
SLAVE TALKER, 23
SM bus, 50, 51, 215
SM- bus, 16
smart batteries, 52
SMbus, 50
SMBUs, 91
SMLINK1, 90
Special addresses, 44
Speed, 47

stalling, 32
Standard, 47
start, 27
Start, 45
START, 24, 25, 26, 27, 28, 29, 35, 36, 37, 39, 40, 42, 45, 69, 70, 72, 77, 78, 80, 81, 83, 136, 167, 215, 229
Stop, 27
STOP, 24, 25, 26, 27, 28, 31, 32, 35, 36, 37, 38, 39, 40, 41, 42, 43, 45, 46, 47, 49, 70, 72, 74, 77, 78, 126, 136, 137, 138, 139, 151, 153, 178, 220
Successive approximation, 108
summing DAC, 107
Switched Resistor Network, 106
synchronisation, 42
TDA7561, 115
test, 73

Thermometer DAC, 105
TLC59116, 102
TotalPhase, 71
trigger, 80
TV400, 82, 83
UL1577, 63
ULN2803, 185, 189
USB, 17, 57, 72, 75, 82, 90, 132, 134, 135, 141, 144, 240
USB2.0, 72
USBee, 71, 72
Video, 115
Windows, 90, 91
Windows 95, 90
WRITE, 29, 30, 35, 37, 38, 78, 79, 149, 165
WRITE address, 30
Writing, 35

Board 1

This is the base board and contains a basic set of I²C devices. It also contains a USB to I²C adapter that lets you excercise the devices. A software platform is available for download that gives you an abstraction layer to all the submodules.

The board is a panel with breakaway tabs. You can simply 'snap' the individual modules apart. The bus connector is present twice for every module. In this way you can simply click together the sub units of interest and try out your bus configuration before you make your own board. Every module has solder bridges that let you program the address of the devices on the unit. Devices that have interrupt capability can be connected, or left disconnected, from the interrupt line. The board has the following sub modules:

USB to I²C Interface
This board forms a bridge between a PC application and the USB bus. It has programmable clock speed and uses a human readable command set. There are no drivers required as it enumerates as an HID device. On board LEDs indicate the bus status and device status. Extra outputs are available that supply logic signals depending on the bus state. These are useful if you want to trigger an oscilloscope or logic analyser on a specific event such as START, STOP, ADDRESS ACK and NORMAL ACK.

PCA 9534 Protected Input
The protected input module features a PCA9534 device and has 8 inputs that can withstand up to 24 volts. The inputs are NOT opto isolated. Each input has its own LED indicating input state.
The PCA9534 was chosen over the classic PCF8574 because it features mask registers and a, pin by pin, programmable input polarity. Connections to the board use a screw terminal block.

PCA8563 Realtime Clock with Battery Backup
In an embedded system it is frequently necessary to have access to accurate time and date. This realtime clock module with backup can provide exactly that functionality. The PCA8563 has a programmable alarm output, or can provide an accurate heart-beat signal to your system. Time, date and day of the week are automatically tracked. The board uses a 'supercap' to provide power when the main system is shut down. The capacitor is good for a few months of non stop operation and charges in seconds. Unlike batteries it does not wear out or needs replacement.

PCA 9534 Protected Output
The Protected Output is the counterpart to the Protected Input board. It has an 8 channel output through an ULN2803 device. The controller, just like the protected input, is also a PCA9534. Off board connections are through a screw terminal block.

LM75 Temperature Sensor
The LM75 temperature sensor is an all in one temperature sensor and monitor. It can be programmed to transmit a logic signal when certain, user specified, thresholds are passed. This board allows for the easy integration of this sensor in your I²C bus system.

LCD and Keyboard Module
This module contains a 2x8 character LCD display and 16 keys. The module uses some clever logic and a matrix configuration of the keys to be able to perform its functionality. It does require some overhead as it is the bus controller that needs to interrogate and scan the keyboard. This is an advanced module that can show you how to do quite a few operations with a single I/O device on the I²C bus.

Bus Power Supply
This experimentation board also hold a little power supply module. It uses the classic 7805 regulator to provide all the boards with power. Input can be between 7 and 9 volts without the need for a heatsink. Up to 15 volts can be handled by attaching a small heatsink to the regulator. Just like its real I²C counterparts, this board also has the double bus connector so it can be placed anywhere in the system.

24xxx EEPROM Module
The Eeprom module can host any of the 24Cxx or 24cxxx devices as well as the PCF8581 and PCF8582 EEPROMs. For devices that have this capability, the board has a bridge to block write access to the memory array.

PCA 9553 PWM LED Controller
This module uses a PCA9553 chip to drive 2 LEDs: one RGB and one white. The intensity of each channel can be programmed independently through the PWM registers. An on chip second oscillator allows for automatic blinking of each of the channels.

Further information and ordering: www.elektor.com/labworx

Board 2

This board complements board #1 and offers even more modules with various functions. Some modules, like the protected input and output, are repeated. The idea is to make the I/O robust so it will survive experimentation on the bench. In a typical I²C implementation there are only a few memories or special functions while I/O remains the main application. Just like its counterpart this board is a panel with breakaway tabs. You can simply 'snap' the individual modules apart. The bus connector is present twice for every module. In this way you can simply click together the sub units of interest and try out your bus configuration before you make your own board.

Every module has solder bridges that let you program the address of the devices on the unit. Devices that have interrupt capability can be connected, or left disconnected, from the interrupt line. The board has the following sub modules:

PCA 9534 Protected Input
The protected input module features a PCA9534 device and has 8 inputs that can withstand up to 24 volts. The inputs are NOT opto isolated. Each input has its own LED indicating input state. The PCA9534 was chosen over the classic PCF8574 because it features mask registers and a, pin by pin, programmable input polarity. Connections to the board use a screw terminal block.

PCA 9534 Protected Output
The protected output is the counterpart to the Protected Input board. It has an 8 channel output through an ULN2803 device. The controller, just like the protected input, is also a PCA9534.
Off board connections are through a screw terminal block.

PCF8574 I/O Board
This is a simple I/O board that has no additional drivers or input protection. It can handle a variety of I/O chips such as the PCF8574, PCF8574A, PCA9534, PCA9554 and a few more. As usual, device addresses and interrupt capability may be set by solder bridges.

SAA1064 LED Display
The SAA1064 is an I²C compatible chip that can drive up to 32 individual LEDs. LED drive current is software controlled. In this module the SAA1064 is set up to drive four 7-segment displays. Ideal to display a numerical value that needs to be visible from a distance.

PCA9544 Bus Expander
The PCA9544 bus expander allows you to partition an I²C bus. Think of it as a router. Just like in a computer network, this device can combine multiple I²C buses into one. If you run out of address space in one branch, simply continue in the other branch.

PCF8591 AD/DA
The PCF8591 is one of the oldest I²C devices out there, but it is still a very useful device. It combines a 4 channel 8 bit analog input with an 8 bit analog output. An instrumentation system allows the inputs to be set up as either single ended or differential.

ADC121 A/D Converter
This module contains a 12 bit A/D converter in SOT23 package. Multiple devices can share the bus. These are ideal to scatter around a system since you can get the precise signal digitized on the spot and don't have to carry it far.

MCP40D17 Potentiometer
This module contains a programmable resistor. Through the I²C bus you can set the wiper value. Depending on the chip installed, this value can also be saved in EEPROM to be automatically restored on power up.

MCP4725 D/A Converter
This forms the counterpart to the 12 bit A/D module. It has 12 bit DACs in a SOT23 package. This module lets you play with these interesting devices.

24xxx EEPROM Module
The EEPROM module can host any of the 24Cxx or 24xxx devices as well as the PCF8581 and PCF8582 EEPROMs. For devices that have this capability, the board has a bridge to block write access to the memory array.

LM75 Temperature Sensor
The LM75 temperature sensor is an all in one temperature sensor and monitor. It can be programmed to transmit a logic signal when certain, user specified, thresholds are exceeded. This board allows for the easy integration of this sensor in your I²C bus system.

Further information and ordering: www.elektor.com/labworx